T0214741

HOW PRESCRIPTION AND OVER-THE-COUNTER DRUGS AFFECT SEXUAL PERFORMANCE

HOW PRESCRIPTION AND OVER-THE-COUNTER DRUGS AFFECT SEXUAL PERFORMANCE

Robert B. Raffa
Patricia J. Bush
Albert I. Wertheimer

CRC Press
Taylor & Francis Group
Boca Raton London New York

CRC Press is an imprint of the
Taylor & Francis Group, an **informa** business

First edition published 2021
by CRC Press
6000 Broken Sound Parkway NW, Suite 300, Boca Raton, FL
33487-2742

and by CRC Press
2 Park Square, Milton Park, Abingdon, Oxon, OX14 4RN

© 2021 Taylor & Francis Group, LLC

CRC Press is an imprint of Taylor & Francis Group, LLC

The right of **Robert B. Raffa, Patricia Bush, and Albert I.
Wertheimer** to be identified as the authors of the editorial material,
and of the authors for their individual chapters, has been
asserted in accordance with sections 77 and 78 of the Copyright,
Designs and Patents Act 1988.

Reasonable efforts have been made to publish reliable data and
information, but the author and publisher cannot assume
responsibility for the validity of all materials or the consequences of
their use. The authors and publishers have attempted to trace the
copyright holders of all material reproduced in this publication
and apologize to copyright holders if permission to publish in this
form has not been obtained. If any copyright material has not been
acknowledged please write and let us know so we may rectify in
any future reprint.

Library of Congress Cataloging-in-Publication Data

ISBN: 978-0-367-49059-1 (hbk)
ISBN: 978-0-367-49057-7 (pbk)
ISBN: 978-1-003-04426-0 (ebk)

Typeset in Palatino
by MPS Limited, Dehradun

Table of Contents

Preface

If you inquire, you are likely to find that many of your patients, colleagues, or anyone else that is professional enough or brave enough to answer are not satisfied with their sex lives. In 2015 we published *Your Drugs & Sex*, a paperback intended for the average person who was not satisfied with their sex life and wanted to know if a medicine could be to blame—and if yes, what to do. We learned that many people were unhappy but were too afraid to ask their healthcare providers—and just as importantly, that most healthcare providers did not ask. The fact that prescription and over-the-counter drugs can affect sexual performance is not often considered. For example, many older people and their providers simply assume that a decline in sexual performance is just an inevitable consequence of aging. The reality is that some drugs are able to improve sexual problems, while others exacerbate them. We believe that healthcare providers have an opportunity to greatly enhance their patients' quality of life by having a basic knowledge of the major drug categories and individual drugs that are important for this topic.

Many new drugs, both prescription and over the counter, have entered the pharmaceutical market since 2015, and more science has begun to address drugs and sex. The purpose of this book is to provide all healthcare professionals contemporary information, the sources of that information, and methods to treat their patients. In almost all situations where a pharmaceutical is causing a sexual performance issue, there is an alternative therapy that does not contribute to the problem. We used information published in respected, peer-reviewed journals. We are aware that there are advertisements in the lay press for products claiming to work wonders, but we have omitted mention of products that do not have verified studies supporting their claims.

We have striven to be sensitive to, and inclusive of, individual differences, lifestyles, and sexual orientations and practices, and to evolving terminology. But we are aware that we're likely to have misstepped on occasion, and apologize for any unintended slights. And we would like to point out that much more scientific study is clearly needed.

We sincerely thank Jo Ann LeQuang for her incredible writing skills, without which this book would still be in the planning phase. We also sincerely thank Hilary Lafoe and Jessica Poile of Taylor & Francis for their professionalism and encouragement and support throughout the very pleasant process of getting this book started, then completed!

The Authors
Robert B. Raffa, PhD, Tucson, AZ
Patricia J. Bush, PhD, Naples, FL
Albert I. Wertheimer, PhD, Naples, FL

About the authors

Robert B. Raffa, PhD is an adjunct professor at the University of Arizona College of Pharmacy (Tucson, AZ) and professor emeritus at Temple University School of Pharmacy (Philadelphia, PA). He has bachelor's degrees in chemical engineering and physiological psychology, master's degrees in biomedical engineering and toxicology, and a doctorate in pharmacology. He was a research fellow and team co-leader for analgesics drug discovery at Johnson & Johnson. He was a full professor at Temple University School of Pharmacy. He is currently a consultant, a cofounder of CaRafe Drug Innovation and Enalare Therapeutics, and the CSO of Neumentum Inc. He is a coholder of several patents and has published more than 350 papers and coauthored or edited several books. He is a past president of the Mid-Atlantic Pharmacology Society of the American Society of Experimental Therapeutics and a recipient of National Institutes of Health funding, and research and teaching awards. He currently lectures and consults worldwide on pharmacology principles and drug discovery and development.

Patricia J. Bush, PhD, is an emeritus professor at Georgetown University School of Medicine, where she chaired the Division of Children's Health Promotion in the Department of Family Medicine. She earned a BS in Pharmacy from the University of Michigan, an MSci in medical sociology from the University of London (UK), and a PhD in social pharmacy from the University of Minnesota. She has received numerous research grants and has published eight books, 21 book chapters, and more than 110 articles in peer-reviewed journals. She has done research in other countries including Moldova. She has been a consultant to a number of organizations, including the US Pharmacopeia. Her areas of expertise include medicine-use behaviors, and she is best known for her studies in children's medicine knowledge, attitudes, and behaviors. Her awards include the 2019 Distinguished Alumni Lifetime Award from the University of Michigan College of Pharmacy.

Albert I. Wertheimer, PhD, MBA, is a professor at the College of Pharmacy, Nova Southeastern University, in Ft. Lauderdale, Florida. He is the founding editor of the *Journal of Pharmaceutical Health Services Research*, published by Oxford University Press, and an international authority in pharmacoeconomics and outcomes research. He is an author or editor of 43 books and author or coauthor of more than 400 professional journal articles. He initiated and chaired the Department of Social and Administrative Pharmacy at the University of Minnesota. He has been the recipient of numerous awards and honors and has lectured or consulted in over 75 countries.

1 Anatomy, Physiology, Arousal, Orgasm, and Hormones

1.1 INTRODUCTION: BACKGROUND AND CONTEXT

Human sexuality has inspired writers, musicians, painters, sculptors, poets, and other artists over the centuries (Fig. 1.1), but a complete understanding of the scientific basis of its functioning—and dysfunctioning—continues to elude scientists. While human sexuality remains a fundamental cornerstone of a happy, healthy life, what drives sexual response and contributes to sexual satisfaction has not been thoroughly elucidated. In part, this is because human sexuality involves the complex interplay of cerebral impulses, emotional responses, psychological attitudes, societal and religious mores, and physical behaviors; in part, it is because human sexuality is hard to quantify and until recently was rarely discussed as a legitimate scientific topic. The more we learn about the brain and its chemicals, the more we find that they drive sexual response, as do hormones and neurons and previous learned responses. Practically anything can affect human sexuality, including age, health, culture, and prior experiences. Things like stress and despair can defeat our sexual impulses, while the mysterious process of falling in love can rapidly sharpen sexual desires. Among the many things that can affect sexual function are drugs, and to better understand how drugs can change our sexual responses, it is necessary to have a thorough overview of what goes on in human sexuality.

1.2 ANATOMY AND PHYSIOLOGY

The penis and the clitoris both contain a dense network of highly sensitive nerves and are very susceptible to sexual stimulation. The penis (Fig. 1.2) is filled with spongy tissue that becomes engorged with blood when sexually aroused, resulting in an erection (Fig. 1.3). This is possible because the skin of the penis is loose enough to accommodate what may be a substantial change in size and shape (Fig. 1.4). During an erection, the urethra (which normally carries urine) is blocked so that only semen can be expelled during orgasm. Sperm is produced in the two testicles (testes) contained in the saclike structure of the scrotum (Fig. 1.5). It is made in the seminiferous tubules within the testicles (Fig. 1.6) and then travels to the epididymis, where it is stored awaiting sexual arousal; this creates contractions that move the sperm from the epididymis into the vas deferens and then to the urethra, from which it is ejaculated (Fig. 1.7). Seminal vesicles are little pouches near the vas deferens at the base of the bladder that produce a fluid that gives the sperm motility. Most of the ejaculate material is actually seminal fluid. The fluid travels through the urethra, which takes it through the center of the prostate gland. The prostate gland is just below the bladder and produces fluid to aid in ejaculation. Most commonly, orgasm occurs with ejaculation.

The vulva is an external sexual organ (Fig. 1.8). The labia majora and labia minora cover the vagina, which is located below the urethral opening (Fig. 1.9). At the top of the vulva, where the labia minora converge, there is a small spongy organ called the clitoris that is densely innervated and highly responsive to stimulation. It is made of spongy tissue that engorges with blood and becomes erect during sexual arousal. Although often compared to the penis because of its spongy tissue and sexual responsiveness, the clitoris holds the unique distinction of being the only human organ exclusively for the purpose of sexual pleasure. The vagina is a passageway that connects the external sexual organs to the

1

Figure 1.1 Human sexuality expressed in works of art.

uterus or womb (Fig. 1.10). The entrance to the uterus is called the cervix. The vagina may be thought of as folds of winkled skin about two to four inches long. When the woman is aroused, the vagina becomes engorged with blood and may extend to double its length or more as it expands to accommodate a penis.

The cervix or neck of the uterus is normally open; it is the passage through which menstrual blood sloughed off from the uterus passes on its way through the vagina and out of the body. When a person is pregnant, the cervix tightens to protect the uterus, but it expands, or dilates, when childbirth is imminent. During most deliverie, the cervix stretches to maximal capacity to permit the baby to exit the womb and enter the birth canal (Fig. 1.11). The uterus is a pear-shaped organ about the size of afist that stretches many times its original size to accommodate a baby during pregnancy (Fig. 1.12). When sexually aroused, the lower end of the uterus tilts toward the abdomen, providing extra space in the vagina for intercourse, in a process known as "tenting."

Eggs are produced in ovaries (Fig. 1.13), two organs that also produce the hormones estrogen (estradiol), progesterone, and testosterone (Fig. 1.14). Eggs

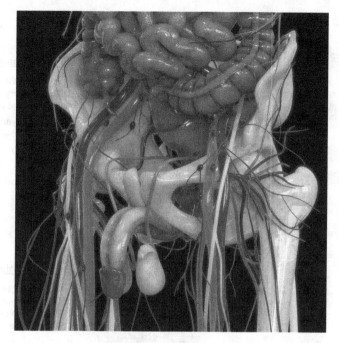

Figure 1.2 Anatomical location of the penis.

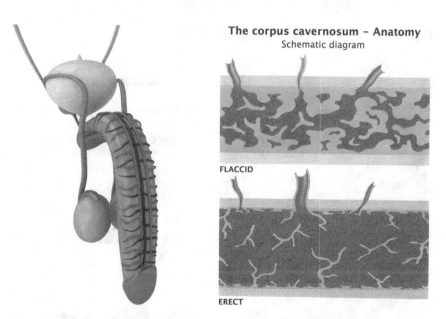

Figure 1.3 Anatomy of penile erection.

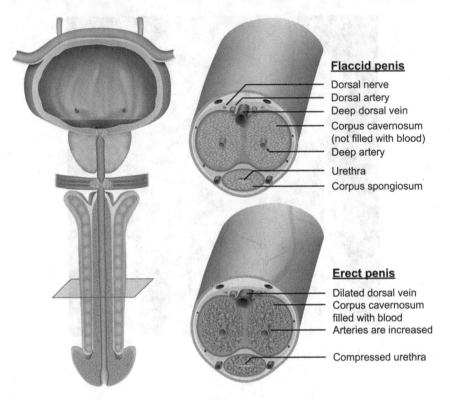

Figure 1.4 Blood supply during penile erection.

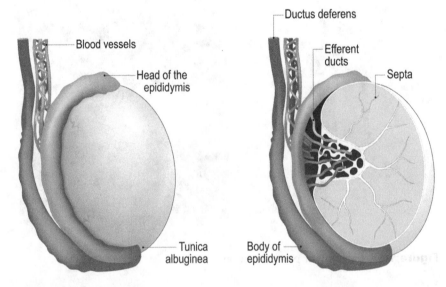

Figure 1.5 The scrotum.

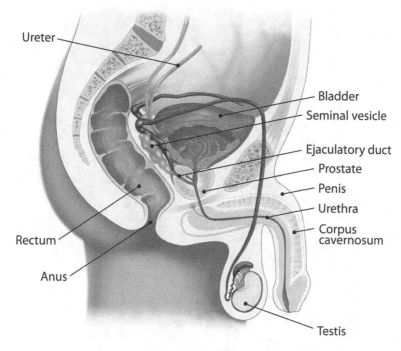

Figure 1.6 Gross anatomy of sperm production.

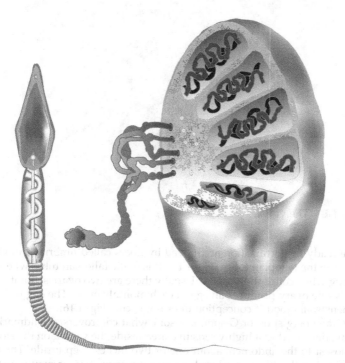

Figure 1.7 Sperm and testicle.

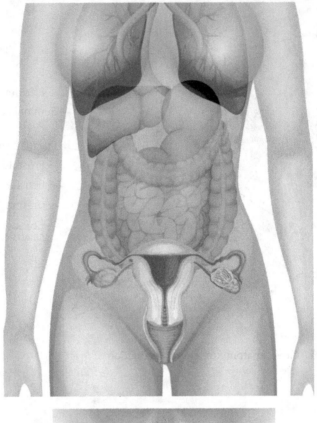

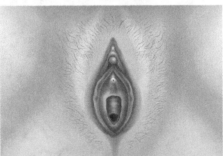

Figure 1.8 The vulva.

that are ready for fertilization are moved by fibers called fimbriae from the ovaries into the fallopian tubes (Fig. 1.15). It is in the fallopian tubes where sperm meets egg when conception occurs. Usually there are two ovaries, but it is typical that only one ovary produces an egg in each monthly cycle. The egg is discarded with menstrual blood if conception does not occur (Fig. 1.16).

The Gräfenberg spot, or G-spot, is a somewhat controversial landmark in anatomy, described as a highly sensitive area inside the vagina on the side of the body closest to the abdomen, about one or two inches deep inside. The G-spot, first defined in the 1950s as a highly erogenous zone, may or may not even

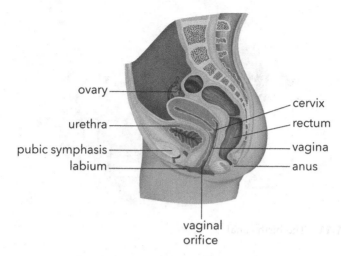

Figure 1.9 Anatomical relationship of female sexual organs.

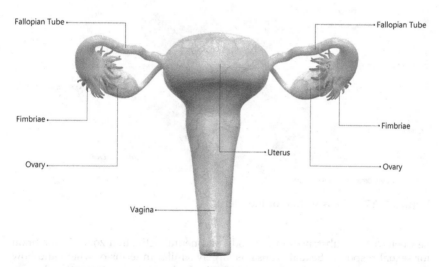

Figure 1.10 Gross anatomy of female reproductive organs.

exist.[1,2] However, despite this controversy, the search to prove or refute the existence of the G-spot has led to considerable advances in our understanding of sexuality.[3]

1.3 AROUSAL

Arousal is a psychological, emotional, and physiological response to sexual stimuli that can be grouped into "reactive arousal," which occurs in direct response to physical stimulation or a sexual encounter, and "spontaneous arousal," which occurs in response to a person's affective thoughts or fantasies. Arousal has a strong biological framework, in that it involves the interplay of

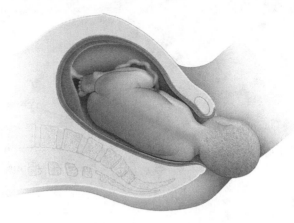

Figure 1.11 The birth canal.

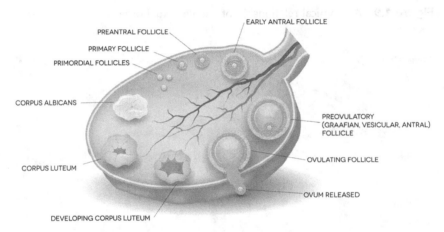

Figure 1.12 Cross section of the uterus.

neuroendocrine substances in the body and neural activation zones in the brain for sexual response. Sexual arousal is largely similar in terms of where and how the brain responds to sexual cues, but individuals can differ in emotional response to sexual stimulation. The neural pathways for sexual stimulation involve the sympathetic and parasympathetic nervous systems and the spinal cord (Fig. 1.17). The pudendal nerve transmits sexual arousal in the brain to the genitalia. Sexual stimulation increases blood flow to the genitals, resulting in penile erection and subsequent ejaculation with orgasm. The increased blood flow to the genitals results in erection of the clitoris and vulvar hyperemia, but may or may not culminate in orgasm.[4]

In terms of brain chemicals, sexual arousal involves the interplay of multiple neurotransmitters, including dopamine, oxytocin, and serotonin (Fig. 1.18), nitric oxide, noradrenaline (norepinephrine),vasoactive intestinal peptide, and neuropeptide Y.[4]

Despite the importance of this subject, remarkably little is known about the neurological processes involved in sexual arousal and response. There are

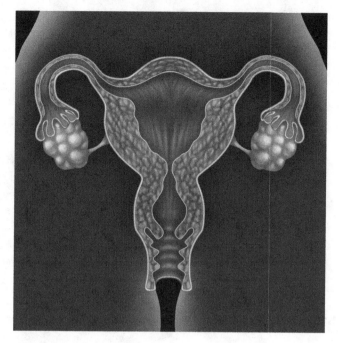

Figure 1.13 Egg production in ovaries.

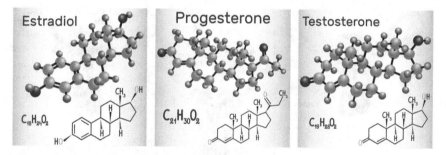

Figure 1.14 Chemical structures of estradiol, progesterone, and testosterone.

psychosomatic frameworks that humans use to find and attract willing sexual partners; emotional, social, physiological, and chemical interplay between the individuals prior to the sexual act; and then the sexual act itself, under autonomic modulation by sexual hormones. For example, chemical mediators known as pheromones work on a prospective partner by being transmitted to their central nervous system.

Thus, sexual arousal involves the interaction of the body's somatic and autonomic systems in ways that blur the boundaries between them. The somatic aspects of the sex act are our response to an attractive potential partner, a desire for pleasure, and the goal of achieving sexual satisfaction. In broad terms, the somatic system regulates how a human interacts with their environment and other people. The autonomic nervous system regulates the body's control over its

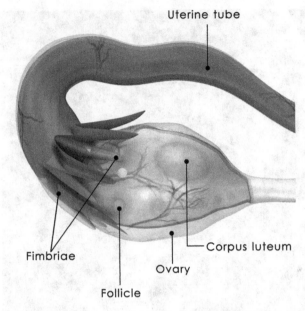

Figure 1.15 Ovary, fimbriae, and fallopian (uterine) tube.

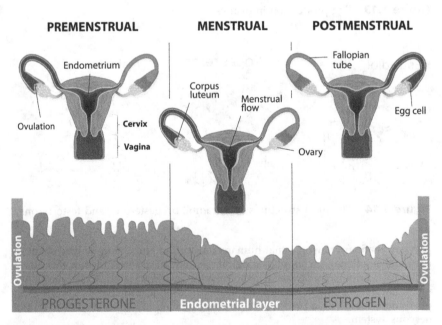

Figure 1.16 Hormone levels and anatomical changes during the menstrual cycle.

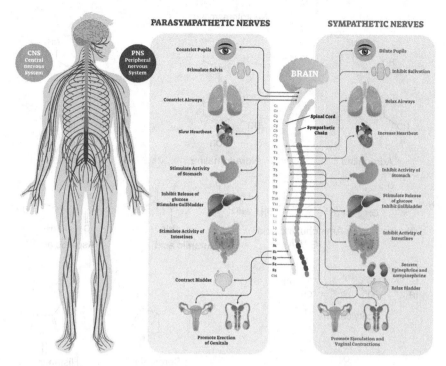

Figure 1.17 The subdivisions of the autonomic nervous system.

own internal organs. Autonomic drivers for sexual activity may be a natural impulse to procreate or a desire to experience orgasm for overall wellness and health. While it is known that pheromones exist and play a role in sexuality, it is not known whether they help stimulate the somatic or autonomic aspects of sexual arousal, whether they are even necessary for sexual arousal, and whether they affect people in different ways.[5] Two pheromones are produce in sweat—androstenol and androstenone (Fig. 1.19)—and these have been shown to affect libido, ovulation, and menstruation. Exposure to high levels of these pheromones tends to result in higher levels of cortisol. Other pheromones, copulins, are produced by the vagina and are known to affect perceptions of others as possible mates.[6] One study found that men exposed to the scent of women near ovulation had higher testosterone levels than men exposed to the scent of nonovulating women or no scent at all.[7]

The dual theory of sexuality maintains that sexual arousal causes neurotransmission to the spine and that this messaging then ascends upward to the brain (afferent pathways) and descends back down to the genital region. This spinal passage of sexual desire occurs regardless of whether sexual stimulation is physical or affective. The brain is intensely involved in sexual response; various cerebral areas coordinate a complex interplay of physiological responses to sexual stimulation. In the male physiology, these responses are erection, preejaculatory lubrication of the penis, swelling of the testicles, thickening of the scrotum, tachycardia, and rapid breathing.[8] In the female physiology, the vagina becomes lubricated, the cervix elevates, the uterus is displaced to expand the vaginal vault, the breasts swell, the nipples become erect, and the clitoris becomes erect.[9]

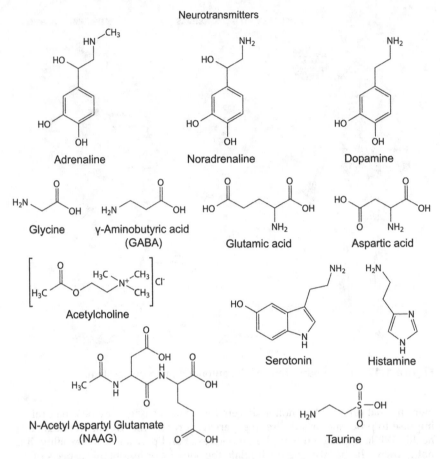

Figure 1.18 Chemical structures of important neurotransmitters.

1.4 ORGASM

In the animal kingdom orgasm is closely related to ejaculation and reproductive success, and there are strong selective pressures to assure regular orgasms with sexual encounters, whereas femaleorgasms, like female sexuality, are more complicated and variable.[10] It is thought that 10% of people do not experience orgasm at all from any sort of sexual activity.[11] The differences in rates of orgasm have been attributed to evolution, culture, societal influences, and anatomical features, but no convincing evidence in support of any particular argument or interplay of arguments has been made in the medical literature.[10]

Penis orgasm, currently understood as emission and ejaculation, may be vastly more complex than that. Afferent stimulation that leads to erection is the same as or similar to that which produces an ejaculation. It has been speculated under the dual cerebral theory of sexuality that ejaculation is actually a two-phase process which begins in the sympathetic and parasympathetic nervous systems of the brain. We can distinguish between the cerebral process of ejaculation and the spinal process of ejaculation, in that sexual arousal and erection are afferent processes in the orgasm, while the actual ejaculation is an efferent or spinal

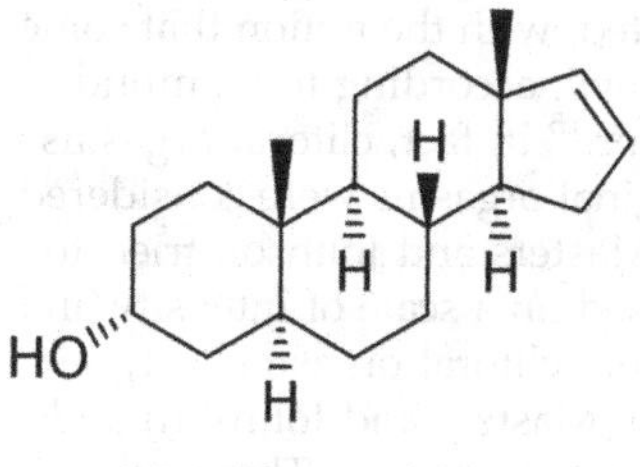

Androstenol

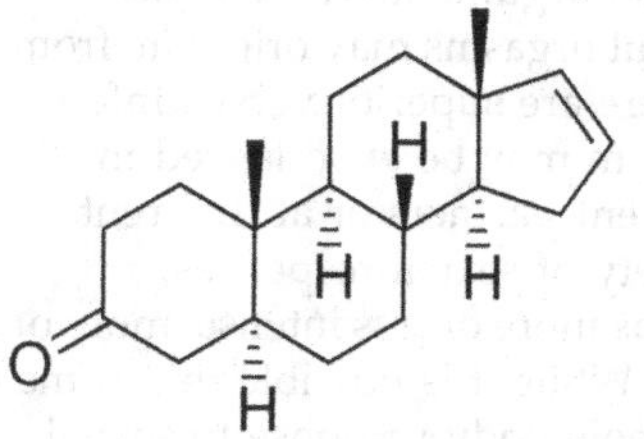

Androstenone

Figure 1.19 Chemical structures of androstenol and androstenone.

process. Emission is the first phase of the orgasm, and it launches the sequence that culminates in orgasm. Somatic impulses travel afferent pathways to the spine (T10 to L3 levels). This launches autonomic efferent messaging to help contract the smooth muscles of the prostate, vas deferens, and seminal vesicles. Some of the seminal fluid is excreted at this point into the posterior urethra, the bladder neck contracts, and then emission occurs. The second phase of ejaculation is sometimes termed expulsion, as it involves the deposit of semen in the urethra, which is then forcefully expelled by pulsatile muscular contractions.[12] It is this latter phase that is more associated with the pleasurable release experienced during orgasm. There is a refractory period following ejaculation characterized by lack of erection, loss of interest in sexual activity, and a satisfied feeling of relaxation. The neurotransmitters involved at this time are oxytocin and prolactin, both of which enhance mood and positive emotions. The duration of this refractory period may range from under an hour (for younger, stronger persons) to hours or even days in older persons.

In the female physiology, orgasm occurs with rhythmic contractions of the perineal muscles. Adrenergic and nonadrenergic noncholinergic neurotransmitters play a part in genital response to sexual arousal. Vasoactive peptides and neuropeptides can regulate vascular and nonvascular smooth muscles and epithelium. The hormonal environment can play a role in tissue health and function, and since hormones fluctuate on a cyclic basis as well as over a lifetime, there is more variable sexual responses. For instance, with an estrogen deficiency, which may be iatrogenic or may be a normal age-related process, there is lower expression of sex steroid receptors and less lubrication and blood flow to the genitalia.[13] But the picture is even more complex: Endothelial nitric oxide production appears to play an important role in regulating the structures of the sexual organs, but this role has not been completely elucidated.[14]

Of course, the whole notion of 'female' orgasm has been the subject of study, controversy, and dispute for centuries. Orgasm is not required for reproduction. It is perfectly possible for an anorgasmic woman to conceive and bear children.

Even the source of stimulation has also been disputed, with the notion that some orgasms are clitoral in origin while others are vaginal; according to Sigmund Freud, vaginal orgasms are psychologically superior.[15] In fact, clitoral orgasms were considered infantile and immature, while vaginal orgasms were considered better for an emotionally developed adult. Later, Masters and Johnson tried to say there were three different types of orgasms based on a scale of intensity and duration.[16] Proponents favoring the existence of only clitoral orgasms[17] argue that the vagina is largely insensitive (also stated by Masters and Johnson) and that there is nothing in the vagina that would trigger an orgasm. Those who advocate that vaginal orgasms exist as a separate type of orgasm generally argue that vaginal orgasms are psychologically more fulfilling. But this is more of a sociopolitical construct than an actual physiological argument. There is no physiological or anatomical evidence to suggest that orgasms may originate from different parts of the sexual anatomy, nor that there are superior versus inferior orgasms. It seems reasonable to assume that orgasm may be experienced in different ways and that sexual activity with different partners or at different times and in different places may result in a variety of sexual responses, including orgasms that at times may be perceived as more or less intense, more or less gratifying, and of longer or shorter duration. While it is possible that some may perceive orgasms as vaginal, clitoral, or a whole-body response to sexual stimulation, othersn may perceive them as originating only from the clitoris,[18] these may be distinctions without a difference.

1.5 HORMONES

The main hormones in the male physiology are gonadotropin-releasing hormone, follicle-stimulating hormone, the luteinizing hormone, and, perhaps best known, testosterone. All of these hormones are also part of female physiology, although testosterone in much smaller amounts.

Gonadotropin-releasing hormone is produced and regulated by the hypothalamus, a small region of the brain. One of its roles is to stimulate the pituitary gland in the brain (Fig. 1.20) to produce follicle-stimulating hormone and luteinizing hormone, which both regulate reproductive processes. They are the driving forces behind puberty, and they help stimulate the production of sperm in the testicles and keep testosterone levels high. Testosterone is the hormone responsible for traits such as facial hair, heavier bone mass, muscle mass, voice change, and sex drive. Many things, including age, stress, and environmental factors, can affect testosterone levels, which follow a circadian rhythm in healthy individuals.

Hormonal influences in female physiology are more complicated because of the hormonal fluctuations inherent in the normal menstrual cycle, and in transition from from menstruation to menopause. Steroids produced in the ovaries include estradiol, testosterone, and progesterone, all of which are known to modulate libido. During menopause, the ovaries shut down and these hormones decrease markedly, along with libido. But while sexual interest is influenced by these hormones, steroid hormones do not actually generate or improve sexual desire. Estradiol appears to be the most likely steroid to enhance sexuality, although testosterone is often prescribed off-label to increase libido. In a study of naturally cycling women ($n = 43$) tracking sexual desire over two menstrual cycles, it was found that endogenous estradiol increased markedly with sexual desire, endogenous progesterone was predictive of a disinterest in sexual activity, and endogenous testosterone could not reliably predict women's sexual interest or lack thereof.[19]

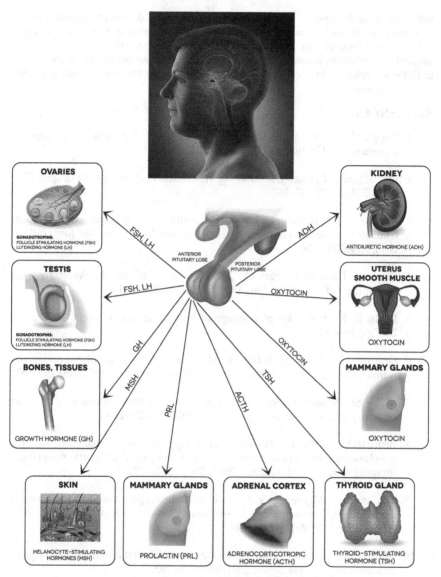

Figure 1.20 The pituitary gland in sexual physiology.

1.6 CONCLUSIONS AND IMPLICATIONS

Sexual libido, arousal, and orgasm are complex cascades of events that require the interplay of various neurotransmitters as well as the synchronized interaction of various somatic, autonomic, peripheral, central, cerebral, and genital functions. Despite the importance of these functions, they have not been thoroughly studied. However, it is apparent that drugs can alter the body's neurochemical balance, which can in turn affect sexuality.

Understanding human sexual activity in terms of neurophysiological processes allows us to better understand how taking drugs can influence sexual desire and behavior. Sexual response is cerebral as well as physical, and is regulated by cerebral neurotransmitters as well as endogenous steroids in a system that has yet to be fully elucidated. Taking prescription drugs, over-the-counter drugs, natural or herbal supplements, or recreational drugs can easily influence this delicate interplay of factors—amplifying desire or disinterest or interfering with sexual response.

REFERENCES

1. Puppo V, Gruenwald I. Does the G-spot exist? A review of the current literature. *Int Urogynecol J*. 2012;23(12):1665–1669.

2. Pan S, Leung C, Shah J, Kilchevsky A. Clinical anatomy of the G-spot. *Clin Anat*. 2015;28(3):363–367.

3. Jannini EA, Buisson O, Rubio-Casillas A. Beyond the G-spot: Clitourethrovaginal complex anatomy in female orgasm. *Nat Rev Urol*. 2014;11(9):531–538.

4. Cour F, Droupy S, Faix A, Methorst C, Giuliano F. Anatomy and physiology of sexuality. *Prog Urol*. 2013;23(9):547–561.

5. Levin R, Riley A. The physiology of human sexual function. *Psychiatry*. 2007;6(3):90–94.

6. Grammer K, Jutte A. Battle of odors: Significance of pheromones for human reproduction. *Gynakol Geburtshilfliche Rundsch*. 1997;37(3):150–153.

7. Miller SL, Maner JK. Scent of a woman: Men's testosterone responses to olfactory ovulation cues. *Psychol Sci*. 2010;21(2):276–283.

8. Brunetti M, Babiloni C, Ferretti A, et al. Hypothalamus, sexual arousal and psychosexual identity in human males: A functional magnetic resonance imaging study. *Eur J Neurosci*. 2008;27(11):2922–2927.

9. Tsujimura A, Miyagawa Y, Fujita K, et al. Brain processing of audiovisual sexual stimuli inducing penile erection: A positron emission tomography study. *J Urol*. 2006;176(2):679–683.

10. Wallen K, Lloyd EA. Female sexual arousal: Genital anatomy and orgasm in intercourse. *Horm Behav*. 2011;59(5):780–792.

11. Wallen K. Commentary on Puts' (2006) review of the case of the female orgasm: Bias in the science of evolution. *Arch Sex Behav*. 2006;35(6):637–639.

12. Motofei IG, Rowland DL. Neurophysiology of the ejaculatory process: Developing perspectives. *BJU Int*. 2005;96(9):1333–1338.

13. Traish AM, Botchevar E, Kim NN. Biochemical factors modulating female genital sexual arousal physiology. *J Sex Med*. 2010;7(9):2925–2946.

14. Musicki B, Liu T, Lagoda GA, Bivalacqua TJ, Strong TD, Burnett AL. Endothelial nitric oxide synthase regulation in female genital tract structures. *J Sex Med.* 2009;6(Suppl 3):247–253.

15. Freud S. *Three essays on the theory of sexuality.* New York: Basic Books; 1962.

16. Masters W, Johnson V. *Human sexual response.* New York: Bantam Books; 1966.

17. Puppo V. Anatomy of the clitoris: Revision and clarifications about the anatomical terms for the clitoris proposed (without scientific bases) by Helen O'Connell, Emmanuele Jannini, and Odile Buisson. *ISRN Obstet Gynecol.* 2011;2011:261464.

18. Pfaus JG, Quintana GR, Mac Cionnaith C, Parada M. The whole versus the sum of some of the parts: Toward resolving the apparent controversy of clitoral versus vaginal orgasms. *Socioaffect Neurosci Psychol.* 2016;6:32578–32578.

19. Cappelletti M, Wallen K. Increasing women's sexual desire: The comparative effectiveness of estrogens and androgens. *Horm Behav.* 2016;78:178–193.

2 Common Drugs and Their Sexual Side Effects

2.1 INTRODUCTION: BACKGROUND AND CONTEXT

The sex system, like any other system in the body, can be affected by pharmacological substances, whether as a drug target or side effect. As age increases, usually so does the use of prescription drugs (Fig. 2.1). This is true for males and females (Fig. 2.2), within racial identities (Fig. 2.3), and within countries by sex (Fig. 2.4) and age (Fig. 2.5). Since the percentage of the world population that is 65 years old or more is rising (Fig. 2.6), and the occurrence of a drug-drug interactions increases with the number of medications taken (Fig. 2.7), pharmacologically induced sexual dysfunction is far from rare. Although sexual adverse effects have become increasingly prevalent, they may be confusing to people who do not realize that the effects may be due to their prescription or nonprescription drugs and instead attribute them to nonpharmacological causes. While many drugs with sexual side effects can be important for a person's overall health, alternative drugs or treatments may be available. The problem is complicated by the fact that many people take drugs long-term for chronic conditions, meaning that sexual side effects that may have been tolerable in the short term become permanent parts of everyday life. The use of over-the-counter products, natural and herbal supplements, and other emerging products such as cannabidiol may also play a role, although people rarely think of these products as a possible cause of sexual side effects.

Healthy sexual function involves neurotransmitters in the brain (Fig. 2.8), including norepinephrine (noradrenaline), dopamine, acetylcholine, gamma-aminobutyric acid (GABA), nitric oxide, oxytocin, serotonin, and others. Arousal and erection are linked to acetylcholine and the parasymphatetic nervous system, while orgasm and ejaculation are likely more controlled by acetylcholine plus norepinephrine in the sympathetic nervous system. As many drugs can affect these neurotransmitters or derange their interplay of these substances, these drugs may result in sexual side effects.[1]

Some of the main classes of drugs that may cause sexual side effects are statins, blood-pressure-lowering drugs, beta-blockers, alpha-blockers, antidepressants, antipsychotics, anxiolytics, antiulcer drugs, and anticonvulsants. Sexual dysfunction associated with drug therapy tends to be more problematic in individuals who had previously enjoyed an active and satisfying sex life rather than older, celibate, or less sexually active individuals.[2] While these effects may be mild and tolerable for some people, they can be profoundly distressing for others. For example, the sexual side effects of certain antidepressants are associated with a worse quality of life.[1]

While older and less sexually active people may take these drugs with little or no problem, there is special concern for adolescents who may take them. First, they may be vulnerable to emotional anguish and pain with sexual dysfunction than older people, as sexual activities are usually new to them. Sexual dysfunction may be devastating to adolescents embarking on explorations of their sexuality. Because of their limited sexual experience, adolescents may be unable to identify or assess sexual side effects.[3]

2.2 STATINS (HMG-COA REDUCTASE INHIBITORS)

Statins, or 3-hydroxy-3-methylglutaryl coenzyme A (HMG-CoA) reductase inhibitors, are a class of drugs taken to lower cholesterol and prevent

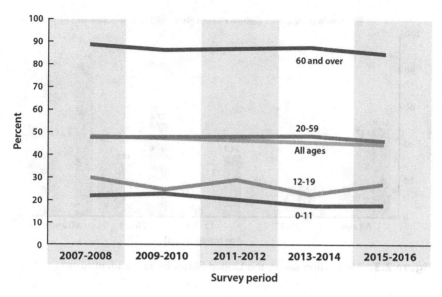

Figure 2.1 Medication use in the United States by age.

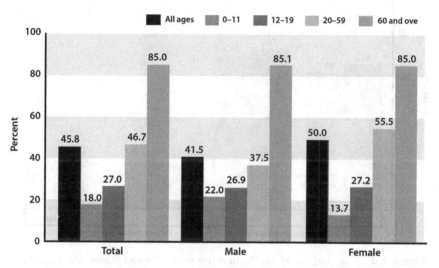

Figure 2.2 Male and female medication use in the United States by age.

cardiovascular disease. Elevated levels of cholesterol in the blood lead to atherosclerosis, which may increase the risk of heart attack, stroke, and peripheral artery disease (Fig. 2.9). Statins bind to and inhibit the activity of HMG-CoA reductase. Human trials using statins such as atorvastatin (Fig. 2.10) have confirmed that changing lipoprotein transport patterns from unhealthy to healthier patterns significantly lowers cardiovascular disease.

While statins are generally well tolerated, there have been reports of muscle pain, behavioral issues, and sexual dysfunction that may be linked to their use.[4]

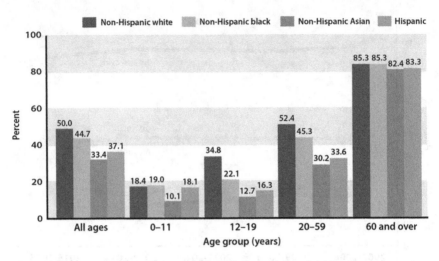

Figure 2.3 Medication use in the United States by racial identity.

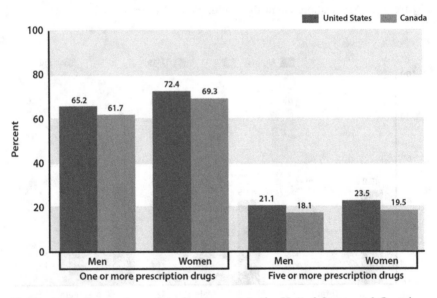

Figure 2.4 Comparison of medication use in the United States and Canada by sex.

Diminished libido and impotence have been reported with statins,[5,6] and three pharmacovigilance databases have reported erectile dysfunction to be associated with statin use.[7-9] This may vary by statin and medical condition. The Scandinavian Simvastatin Survival Study found that simvastatin was no more likely than placebo to be associated with erectile dysfunction in men with coronary heart disease.[10] In contrast, treatment with atorvastatin was associated with improved erectile function in men with hypercholesterolemia.[11] Among people taking statins and experiencing a

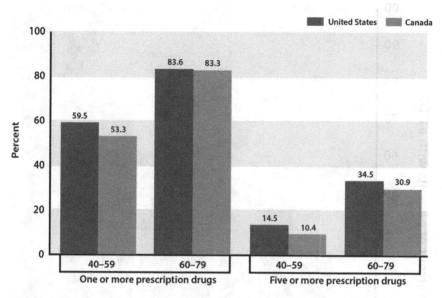

Figure 2.5 Comparison of medication use in the United States and Canada by age.

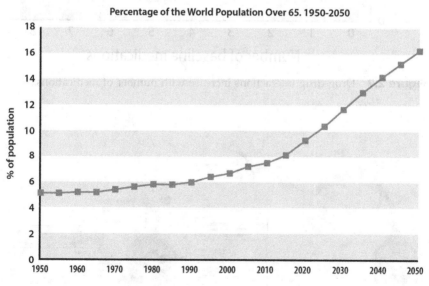

Figure 2.6 Percentage of the world population over 65.

sexual side effect, the median age at onset of symptoms was 53.5 years (range, 40–67 years) with a median onset time of 42 days (range 4–120, days), and recovery time was 14 days (5–35 days). Recovery was based on rotating to a different statin or discontinuing statin therapy.[5,6]

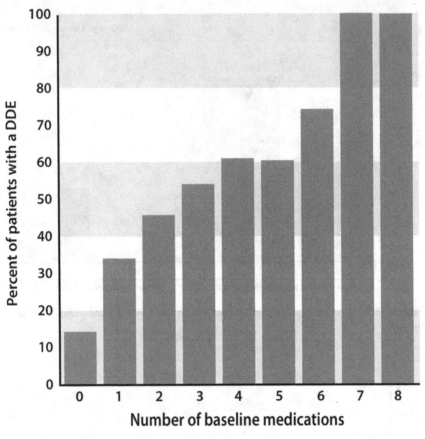

Figure 2.7 Drug-drug interactions increase with number of medications.

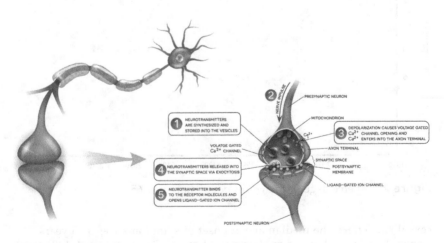

Figure 2.8 Nerve, synapse, and neurotransmitters.

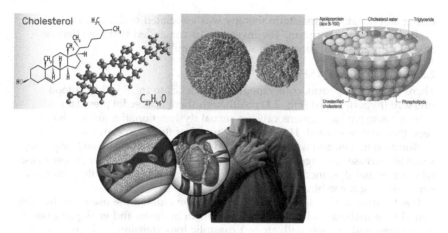

Figure 2.9 Cholesterol and heart attack.

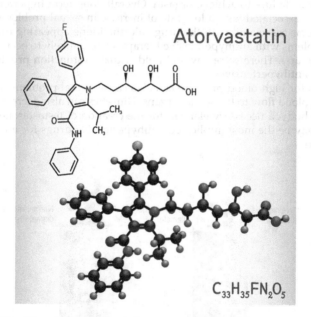

Figure 2.10 Chemical structure of atorvastatin.

It is thought that statins reduce the biosynthesis of sex hormones in the body as they reduce cholesterol synthesis, because cholesterol is a precursor of androstenedione and estradiol. Female sexual dysfunction has not been widely studied with cholesterol use, but one large retrospective cohort study (n = 2,890 women taking statins matched to 2,890 women who did not take statins) found that statins had no association with gonadal or sexual dysfunction, nor were statins associated with menstrual irregularities.[12] In a study of 14 women treated with atorvastatin compared to 14 matched women with normal lipid levels not taking atorvastatin, atorvastatin therapy was associated with decreased sexual desire but increased orgasm. The investigators concluded that the effect of atorvastatin on sexual function in women was mild.[13]

2.3 ANTIHYPERTENSIVES

There are several pharmacotherapeutic approaches to treating high blood pressure (hypertension) (Fig. 2.11). Although there is anecdotal evidence of antihypertensive medications causing sexual dysfunction, this subject has not been thoroughly studied. Hypertension in and of itself can cause sexual dysfunction in men and women, and like decreased sexual desire and response, it tends to increase with age, comorbidities, and stress. It is not clear in some cases whether sexual dysfunction is the result of hypertension, the antihypertensive drug taken, or a combination of the two.

The Treatment of Mild Hypertension Study evaluated the use of acebutolol, amlodipine maleate, chlorthalidone, doxazosin maleate, and enalapril maleate in 902 men and women with stage 1 diastolic hypertension.[14] Upon entry into the study, 14.4% of men and 4.9% of women reported some type of sexual dysfunction. For men, the most common problem was erectile dysfunction; for women it was failure to achieve orgasm. Overall, long-term hypertensive therapy was associated with a low rate of increase in sexual problems for men, with problems associated with the drug chlorthalidone appearing early. Male sexual problems with antihypertensive therapy were unlikely to occur after two years of therapy. There were few and mild sexual dysfunction problems in women on antihypertensives.

Diuretics for high blood pressure can also cause sexual problems by decreasing blood flow to the sexual organs. Diuretics can also decrease levels of zinc in the body, a necessary element for the creation of testosterone. In fact, diuretics may be the most implicated antihypertensive drugs for erectile dysfunction.

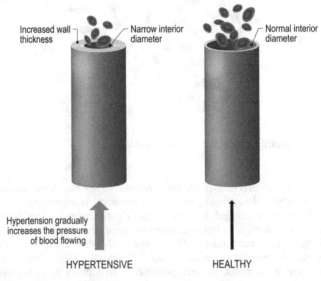

Figure 2.11   Artery internal diameter and hypertension.

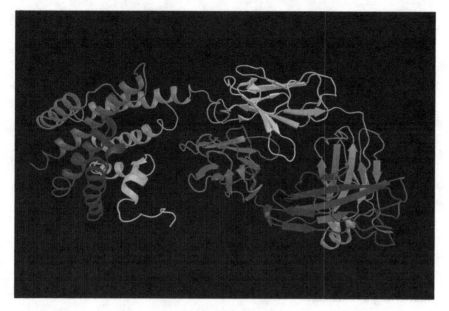

Figure 2.12 "Ribbon" model of beta-adrenoceptor.

Angiotensin-converting enzyme inhibitors, angiotensin II receptor blockers, and calcium-channel blockers are also used to reduce blood pressure but are less associated with sexual side effects.

2.3.1 Beta-Blockers (Beta-Adrenoceptor Antagonists)

Beta-blockers are a class of medications that are used to treat hypertension. They are competitive antagonists that block the receptor sites for the endogenous catecholamines epinephrine (adrenaline) and norepinephrine (noradrenaline) at adrenergic beta-receptors (Fig. 2.12).

The older-generation of beta-blockers, such as propranolol, have been associated with sexual side effects. Captopril and carvedilol, which are newer, are associated with fewer adverse effects of a sexual nature.

2.3.2 Alpha-Blockers (Alpha-Adrenoceptor Antagonists)

Alpha-blockers are drugs used to treat either hypertension or benign prostatic hyperplasia—that is, an enlarged prostate (Fig. 2.13). They are also effective in treating lower urinary tract symptoms. These drugs include alfuzosin, doxazosin, prazosin, silodosin, tamsulosin, and terazosin. They work by keeping norepinephrine from exerting a vasoconstrictive effect on the smaller blood vessels, allowing the vasculature to remain relaxed and conducive to greater blood flow and lower blood pressure. These alpha-blockers work on other muscles as well, and can improve urine flow by relaxing the bladder. All of these drugs are associated with erectile dysfunction, diminished libido, and retrograde ejaculation. Retrograde ejaculation occurs when semen enters the bladder. It does not necessarily prevent orgasm, but is sometimes called a "dry orgasm." Although not harmful, it has been associated with male infertility.

Alpha-blockers often have their most pronounced side effects with the first dose; a commonly reported initial side effect is dizziness, which typically

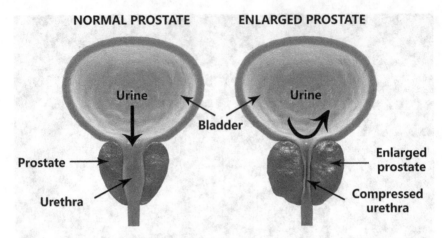

NORMAL PROSTATE   **ENLARGED PROSTATE**

Figure 2.13 Normal and enlarged prostate.

dissipates with long-term use. Alpha-blockers are available in long-acting and short-acting formulations.

2.4 ANTIDEPRESSANTS

Depression often has a negative effect on sexual desire, arousal, and performance, although there is evidence that people who are depressed may still seek and highly value sexual activity.[15] It is estimated that about half of those with untreated depression experience some degree of sexual dysfunction, which can increase with drug therapy for depression.[16] While sexual dysfunction may or may not be a direct result of clinical depression, it may also be caused or exacerbated by antidepressants. In contrast to noradrenergic, dopaminergic, or melatonergic antidepressants, antidepressants that affect the body's serotonergic activity may result in sexual dysfunction. Sexual side effects are not rare; in fact, they occur more frequently than was initially believed. Sexual adverse effects have been reported in 25.8%–80.3% of those taking serotonin reuptake inhibitors (SRIs) and selective serotonin reuptake inhibitors (SSRIs; Fig. 2.14).[17] These antidepressants prevent the reuptake of serotonin in the brain, elevating

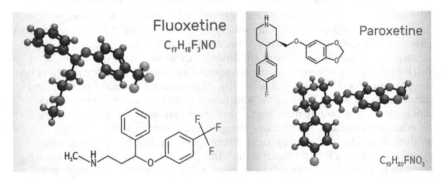

Figure 2.14 Chemical structures of two selective serotonin reuptake inhibitors.

Serotonin pathway

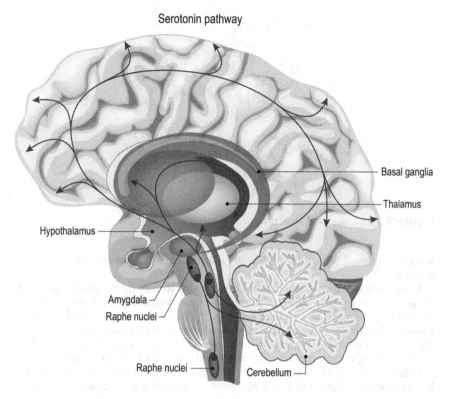

Figure 2.15 Serotonin pathways in the brain.

serotonin levels (Fig. 2.15). In cases of people taking an SRI or SSRI and experiencing sexual problems, 60% involve depressed libido or delayed or no orgasm, and 30% experience difficulty becoming sexually aroused.[2] Erectile dysfunction is reported in 9.5%–34.1% of men.[1] Although men are more frequently affected than women by antidepressants in this way (62.4% vs. 56.9%), women often have more severe dysfunction. Other but less common side effects include priapism, painful ejaculation, loss of vaginal or breast sensation, persistent genital arousal, and nonpuerperal lactation in females.[1] About 40% of people who experience sexual side effects with an SSRI do not find the effect tolerable.[18] In fact, sexual side effects can limit use of or adherence to SRI or SSRI antidepressive treatment.

Serotonin is transported by the monoamine transporter SERT; in general, drugs that block SERT tend to inhibit sexual performance. SERT inhibition may be counteracted by increasing dopamine and noradrenaline.[19] In other words, sexual side effects may occur when SERT is blocked, increasing serotonin neurotransmission.[19] Thus, stimulation of the 5-HT$_2$ receptors may have adverse effects on sexual activity. SSRIs may cause sexual side effects, and it is speculated this is due to inhibition of dopamine signaling pathways.[19] A study of 29 middle-aged men with major depressive disorder (nine taking an SSRI, 10 taking mirtazapine, and 10 healthy controls) found different patterns of brain activation under functional magnetic resonance imaging when the men were provided with neutral and erotic images, suggesting different brain activities in response to

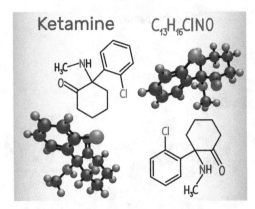

Figure 2.16 Chemical structure of ketamine.

sexual imagery.[20] SRIs and SSRIs affect brain chemistry, and that affects sexuality.

The nature and prevalence of side effects seems to vary by the particular agent. To address sexual side effects, a number of treatment strategies have emerged, including rotating to an antidepressant with a different mechanism of action, adding other drugs (phosphodiesterase inhibitors, amantadine, buspirone, ropinirole, vilazodone, cyproheptadine, or nefazodone), or taking drug holidays such as taking the drug during the week and skipping it on weekends.[1]

Ketamine (Fig. 2.16) offers an exciting new opportunity for novel treatment, but also novel insight into the (patho)physiology of the underlying condition.

2.5 ANTIPSYCHOTICS

Antipsychotic drugs that block dopamine receptors (Fig. 2.17) along dopaminergic pathways in the brain (Fig. 2.18) and increase prolactin (Fig. 2.19) levels may be associated with sexual side effects, to the extent that it compromises drug adherence in younger patients and may lead to discontinuation of treatment.[2] It is estimated that sexual dysfunction occurs in 45%–80% of males and 30%–80% of females on antipsychotic drug therapy. Sexual side effects are generally regarded as the most distressing drug-related adverse event for those with schizophrenia receiving pharmacological treatment.[21] Some antipsychotics have sexual side effects because of histamine receptor antagonism, which increases sedation and in turn diminishes arousal. Some antipsychotics act as dopamine receptor antagonists, which can inhibit motivation and depress reward circuits, in that way decreasing sexual interest, libido, and arousal.[22] Other antipsychotics block the dopamine D_2 receptors along the tuberoinfundibular pathway, which can likewise diminish libido and impair arousal. This mechanism of action can also increase prolactin and, in that way, indirectly impede orgasm. Peripheral vasodilation can be caused by antagonism of either cholinergic receptors or alpha-adrenergic receptors, either of which can cause erectile dysfunction. Antipsychotics have an overall antiadrenergic effect, which may be associated with abnormal ejaculation.[22]

Drugs like haloperidol, risperidone, and amisulpride block D_2 receptors and thus elevate the body's levels of prolactin. Prolactin, naturally secreted by the body, has a normal range of about 10–20 ng/mL in men and 10–25 ng/mL in women. It is highest during the rapid-eye-movement phase of sleep, and levels

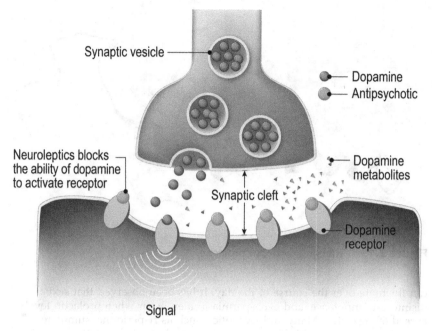

Figure 2.17 Dopaminergic synapse.

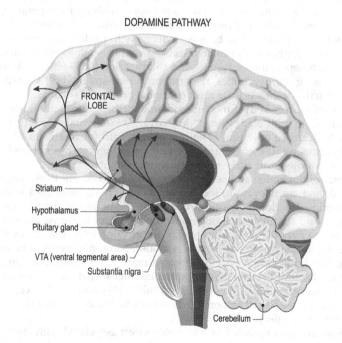

Figure 2.18 Dopamine pathways in the brain.

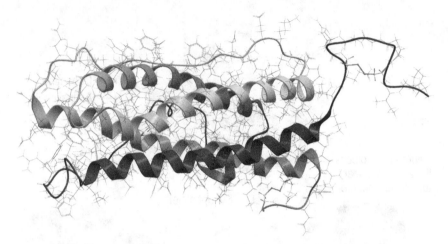

Figure 2.19 "Ribbon" model of prolactin.

can fluctuate over the course of the day. It has been observed that sexual disinterest, impotence, and azoospermia in men occur when prolactin levels exceed 60 ng/mL.[23] Many antipsychotics, such as risperidone, stimulate prolactin synthesis and in that way increase prolactin levels. Other antipsychotics, such as olanzapine, clozapine, quetiapine, ziprasidone, perphenazine, and aripiprazole, are considered prolactin-sparing agents. A prolactin-sparing agent is one that leads to only a modest or temporary increase in prolactin levels, which is contrasted with prolactin-raising agents. Sexual problems have been found to be less frequent with prolactin-sparing drugs (16%–27%) but higher with prolactin-raising agents, such as olanzapine, risperidone, haloperidol, clozapine, and thioridazine (40%–60%).[24] Among the prolactin-raising antipsychotics, risperidone is perhaps the best known for causing sexual side effects.[22] Its most commonly reported sexual side effects are reduced libido, erectile dysfunction, and amenorrhea; rates are higher with risperidone than with olanzapine.[22] Moreover, these sexual side effects do not tend to decrease over time. The role of sildenafil for treatment of sexual dysfunction in men on antipsychotic drug therapy has not been thoroughly studied, but early research is promising.[25]

The roles of dopamine and serotonin in sexual function are increasingly being elucidated, and are relevant to the sexual adverse effects associated with antipsychotics. The dopaminergic structures in the brain are crucial to sexual motivation, and stimulation of dopaminergic receptors in the paraventricular nucleus of the hypothalamus is necessary for erection in males. Drugs that decrease dopamine can impair sexual response. Serotonin seems to be more complex. The serotonergic system is located mainly in the hippocampus and amygdala, and it seems that serotonin inhibits sexual motivation and orgasm. However, stimulation of the serotonin 5-HT$_2$ and 5-HT$_3$ receptors appears to decrease sexual response, whereas stimulation of the 5-HT$_{1A}$ receptors has a stimulatory effect.[23]

Priapism has been reported as a rare side effect associated with atypical antipsychotic drug therapy. This is a painful penile erection not associated with sexual arousal that may persist for very extended periods of time. It is caused

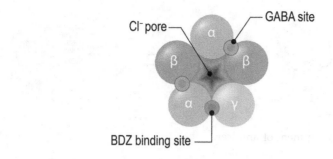

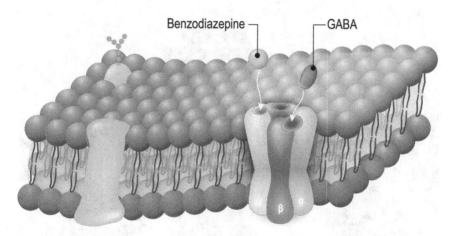

Figure 2.20 The benzodiazepine binding site on the GABA$_A$ receptor complex.

when alpha-adrenergic receptors are blocked. Priapism may be reversible or irreversible, and should be treated as a medical emergency.[23]

2.6 ANXIOLYTICS (ANTIANXIETY DRUGS)

The most commonly prescribed anxiolytics are benzodiazepines (Fig. 2.20), a broad class of drugs that can relieve anxiety and promote sleep. Benzodiazepines are also indicated for certain seizure disorders, spasms and agitation, and alcohol withdrawal syndrome. Because they act as sedatives with strong muscle-relaxing properties, benzodiazepines have also been observed to decrease sexual interest and sexual sensation. They may reduce the production of testosterone, a hormone essential to healthy sexual function in both men and women. The most commonly reported sexual adverse events with benzodiazepines are fewer or less-intense orgasms, anorgasmia, painful intercourse, erectile dysfunction, and problems with ejaculation.

More women than men are prescribed benzodiazepines (about 65%), but less is understood about their sexual side effects in women than in men.

2.7 ANTIULCER DRUGS

Antiulcer drugs (Fig. 2.21) are inhibitors of the gastric proton pump (Fig. 2.22), such as dexlansoprazole, esomeprazole, lansoprazole, omeprazole, pantoprazole, and rabeprazole, which are available both by prescription and some in over-the-

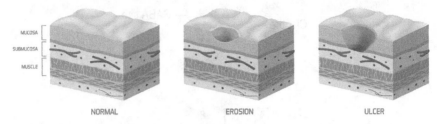

Figure 2.21 Development of an ulcer.

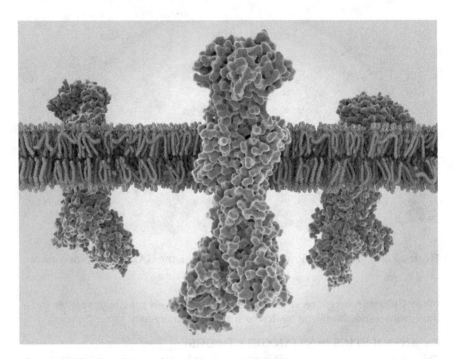

Figure 2.22 Rendition of a gastric proton pump.

counter formulations. It has been theorized that proton-pump inhibitors reduce the free testosterone level and thus set the stage for sexual dysfunction. Since proton-pump inhibitors may be taken in some cases for a short course, sexual problems associated with these drugs have not been widely reported. A variety of adverse events, including a tentative link to dementia, have been reported, but without strong evidence.[26]

2.8 ANTICONVULSANTS

Gabapentinoids, such as pregabalin and gabapentin, are taken to reduce seizures as well as to address neuropathic pain syndromes. These drugs are also used to treat certain mental health conditions. At doses of ≥900 mg/day, gabapentin has been reported to cause sexual side effects up to complete sexual dysfunction. Even at doses of ≥300 mg/day, it has been linked to reduced libido, anejaculation, anorgasmia, and impotence.[27] In a study of 75 people treated with pregabalin for a variety of conditions, sexual dysfunction occurred in 41% and was more common in men than women. The most commonly reported sexual side effects were erectile dysfunction, anorgasmia, and loss of libido. Interestingly, the frequency and severity of sexual side effects were not related to the dose of pregabalin, and the effects could occur within weeks of onset of therapy. These side effects resolved with discontinuation of the drug.[28] Other reports in the literature state that gabapentin-associated sexual dysfunction is dose dependent.[29]

Paradoxically, gabapentin may enhance sexual function in women with vulvodynia, a chronic painful condition that affects the vulva.[30] A case report in the literature describes a 27-year-old man who took pregabalin for depression and developed increased libido and sexual disinhibition.[31] The literature also reports a case of priapism in a 64-year-old man taking pregabalin for about a year for neuropathic pain.[32]

It is thought that gabapentinoids affect areas of the brain involved with sexual arousal, but the exact mechanism of action has not yet been elucidated. Older anticonvulsant drugs such as phenytoin and carbamazepine have a higher association with sexual dysfunction than valproate, lamotrigine, and gabapentinoids. It has been speculated that anticonvulsants, just like epilepsy itself, can cause hormonal alterations that disrupt activity on the hypothalamic-pituitary axis, which in turn may disrupt the production of sex hormones.[33] Further study is needed.

2.9 ANALGESICS

Analgesics can be grouped broadly into opioids (Fig. 2.23) and nonopioids, the latter of which include acetaminophen (called paracetamol in some parts of the

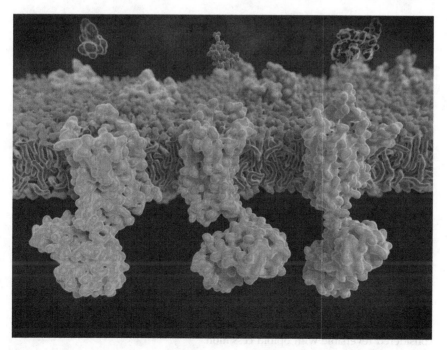

Figure 2.23 Rendition of the opioid receptor types (mu, delta, kappa).

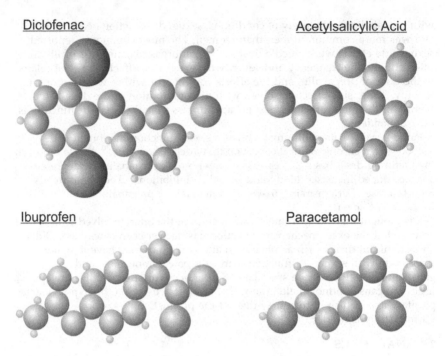

Figure 2.24 Chemical structures of some nonsteroidal anti-inflammatory drugs and of acetaminophen (paracetamol).

world) and nonsteroidal anti-inflammatory drugs (NSAIDs) such as ibuprofen, celecoxib, and naproxen (Fig. 2.24). Despite occasional anecdotal and individual reports of erectile dysfunction associated with NSAID use and one case report in the literature of increased libido with ibuprofen (admittedly self-described as "unusual"),[34] NSAIDs and other nonopioid pain relievers do not appear to be associated with changes in sexual performance or desire, although NSAID use may be associated with gastrointestinal side effects (nonselective NSAIDs) or cardiovascular risk (selective NSAIDs, also known as coxibs). Acetaminophen does not appear to have a proven adverse effect on sexuality.

On the other hand, opioid pain relievers have a long association with adverse sexual effects. Opioids are a broad class of drugs that include morphine, oxycodone, and hydromorphone (Fig. 2.25). As a class, they lower testosterone in men and women, which can reduce libido and sexual desire. They may also cause somnolence and drowsiness, which can diminish sexual interest. These particular effects may be less with buprenorphine than with other opioids, but they can still be considered a class effect.[35,36] Opioids disrupt sex hormones, and they have been associated with hypogonadism in both men and women. Hypogonadism may be defined as gonadal (testicular or ovarian) failure that can directly affect the hypothalamus or pituitary gland. In men, this may result in erectile dysfunction; in women, amenorrhea may develop. Both sexes typically experience a loss of interest in sex with opioid use, but hypogonadism may be more prevalent in men than women. Opioid-induced hypogonadism is considered reversible with opioid cessation.

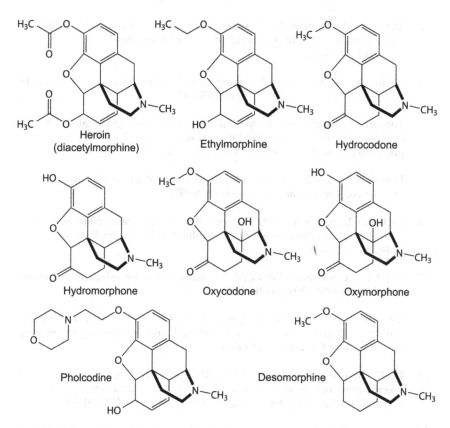

Figure 2.25 Chemical structures of some opioids.

2.10 CONCLUSION

It can be challenging to differentiate sexual problems caused by pharmacological treatments and sexual dysfunction due to underlying health conditions. It is also possible that both conditions contribute to a given sexual dysfunction.

Furthermore, some people find that certain drug therapies improve their overall health and well-being such that they experience greater sexual performance and satisfaction with drug therapy than without it. When treating people with sexual dysfunction, it is important to determine the medications they are taking, including over-the-counter products and supplements. Issues of dose, dependency, and drug-drug interactions must be considered as well as underlying conditions and comorbidities. With sexual dysfunction, it is important to treat a person holistically, as difficult family situations, financial stressors, job-related pressures, and normal aging can also bring about diminished sexual response. In many cases, alternatives to specific drugs may be found which allow a person to resume normal sexual activity without compromising their care.

REFERENCES

1. Francois D, Levin AM, Kutscher EJ, Asemota B. Antidepressant-induced sexual side effects: Incidence, assessment, clinical implications, and management. *Psychiatr Annal.* 2017;47(3):154–160.

2. Montejo LA, Montejo LL, Navarro-Cremades LF. Sexual side-effects of antidepressant and antipsychotic drugs. *Curr Opin Psychiatr.* 2015;28(6):418–423.

3. Levine A, McGlinchey E. Assessing sexual symptoms and side effects in adolescents. *Pediatrics.* 2015;135(4):e815–e817.

4. Tuccori M, Montagnani S, Mantarro S, et al. Neuropsychiatric adverse events associated with statins: Epidemiology, pathophysiology, prevention, and menagaement. *CNS Drugs.* 2014;28(3):249–272.

5. Halkin A, Lossos IS, Mevorach D. HMG-CoA reductase inhibitor-induced impotence. *Ann Pharmacother.* 1996;30(2):192.

6. de Graaf L, Brouwers AH, Diemont WL. Is decreased libido associated with the use of HMG-CoA-reductase inhibitors? *Br J Clin Pharmacol.* 2004;58(3):326–328.

7. Boyd IW. Comment: HMG-CoA reductase inhibitor-induced impotence. *Ann Pharmacother.* 1996;30(10):1199.

8. Carvajal A, Macias D, Sáinz M, et al. HMG CoA reductase inhibitors and impotence: Two case series from the Spanish and French drug monitoring systems. *Drug Saf.* 2006;29(2):143–149.

9. Do C, Huyghe E, Lapeyre-Mestre M, Montastruc JL, Bagheri H. Statins and erectile dysfunction: Results of a case/non-case study using the French Pharmacovigilance System Database. *Drug Saf.* 2009;32(7):591–597.

10. Pedersen TR, Faergeman O. Simvastatin seems unlikely to cause impotence. *BMJ.* 1999;318(7177):192.

11. Saltzman EA, Guay AT, Jacobson J. Improvement in erectile function in men with organic erectile dysfunction by correction of elevated cholesterol levels: A clinical observation. *J Urol.* 2004;172(1):255–258.

12. Ali SK, Reveles KR, Davis R, Mortensen EM, Frei CR, Mansi I. The association of statin use and gonado-sexual function in women: A retrospective cohort analysis. *J Sex Med.* 2015;12(1):83–92.

13. Krysiak R, Drosdzol-Cop A, Skrzypulec-Plinta V, Okopień B. The effect of atorvastatin on sexual function and depressive symptoms in young women with elevated cholesterol levels—a pilot study. *Endokrynol Pol.* 2018;69(6):688–694.

14. Grimm J, RH, Grandits G, Prineas R, et al. Long-term effects on sexual function of five antihypertensive drugs and nutritional hygienic treatment in hypertensive men and women. *Hypertension.* 1997;29:8–14.

15. Gelenberg AJ, Delgado PG, Nurnberg HG. Sexual side effects of antidepressant drugs. *Curr Psychiatr Rep.* 2000;2(3):223–227.

16. Reichenpfader U, Gartlehner G, Morgan LC, et al. Sexual dysfunction associated with second-generation antidepressants in patients with major depressive disorder: Results from a systematic review with network meta-analysis. *Drug Safe.* 2014;37(1):19–31.

17. Serretti A, Chiesa A. Treatment-emergent sexual dysfunction related to antidepressants: A meta-analysis. *J Clin Psychopharmacol.* 2009;29(3):259–266.

18. Montejo AL, Llorca G, Izquierdo JA, Rico-Villademoros F. Incidence of sexual dysfunction associated with antidepressant agents: A prospective multicenter study of 1022 outpatients. Spanish Working Group for the Study of Psychotropic-Related Sexual Dysfunction. *J Clin Psychiatr.* 2001;62(Suppl 3):10–21.

19. Bijlsma EY, Chan JSW, Olivier B, et al. Sexual side effects of serotonergic antidepressants: Mediated by inhibition of serotonin on central dopamine release? *Pharmacol Biochem Behav.* 2014;121:88–101.

20. Kim W, Jin BR, Y'ang WS, et al. Treatment with selective serotonin reuptake inhibitors and mirtapazine results in differential brain activation by visual erotic stimuli in patients with major depressive disorder. *Psychiatry Investig.* 2009;6(2):85–95.

21. Lambert M, Conus P, Eide P, et al. Impact of present and past antipsychotic side effects on attitude toward typical antipsychotic treatment and adherence. *Eur Psychiatr.* 2004;19(7):415–422.

22. Park YW, Kim Y, Lee JH. Antipsychotic-induced sexual dysfunction and its management. *World J Mens Health.* 2012;30(3):153–159.

23. Just M. The influence of atypical antipsychotic drugs on sexual function. *Neuropsychiatr Dis Treatm.* 2015;11:1655–1661.

24. Serretti A, Chiesa A. A meta-analysis of sexual dysfunction in psychiatric patients taking antipsychotics. *Int Clin Psychopharmacol.* 2011;26(3):130–140.

25. Schmidt HM, Hagen M, Kriston L, Soares-Weiser K, Maayan N, Berner MM. Management of sexual dysfunction due to antipsychotic drug therapy. *Cochrane Database Syst Rev.* 2012;11(11):Cd003546.

26. Spechler SJ. Proton pump inhibitors: What the internist needs to know. *Med Clin North Am.* 2019;103(1):1–14.

27. Kaufman KR, Struck PJ. Gabapentin-induced sexual dysfunction. *Epilep Behav.* 2011;21(3):324–326.

28. Hamed SA. Sexual dysfunctions induced by pregabalin. *Clin Neuropharmacol.* 2018;41(4):116–122.

29. Kaufman KR, Struck PJ. Gabapentin-induced sexual dysfunction. *Epilepsy Behav.* 2011;21(3):324–326.

30. Bachmann GA, Brown CS, Phillips NA, et al. Effect of gabapentin on sexual function in vulvodynia: A randomized, placebo-controlled trial. *Am J Obstetr Gynecol*. 2019;220(1):89.e81–89.e88.

31. Murphy R, McGuinness D, Hallahan B. Pregabalin-induced sexual disinhibition. *Ir J Psychol Med*. 2020;37(1):55–58.

32. Karanci Y. Priapism associated with pregabalin. *Am J Emerg Med*. 2019; 38(4):e1-852.e2.

33. Atif M, Sarwar MR, Scahill S. The relationship between epilepsy and sexual dysfunction: A review of the literature. *Springerplus*. 2016;5(1):2070-2070.

34. Kujan O, Raheel SA, Azzeghaiby S, Alqahtani FH, Alshehri M, Taifour S. An unusual side effect of Ibuprofen post dental therapy: Increased erectile and libido activity. *J Int Oral Health*. 2014;6(6):94–95.

35. Bliesener N, Albrecht S, Schwager A, Weckbecker K, Lichtermann D, Klingmüller D. Plasma testosterone and sexual function in men receiving buprenorphine maintenance for opioid dependence. *J Clin Endocrinol Metab*. 2005;90(1):203–206.

36. Hallinan R, Byrne A, Agho K, McMahon C, Tynan P, Attia J. Erectile dysfunction in men receiving methadone and buprenorphine maintenance treatment. *J Sex Med*. 2008;5(3):684–692.

3 Drugs That Treat Sexual Dysfunction

3.1 INTRODUCTION: BACKGROUND AND CONTEXT

The blockbuster drug Viagra (Fig. 3.1) was originally developed in 1989 by Peter Dunn and Albert Wood at Pfizer as an antihypertensive that could possibly treat angina, or chest pain, which occurs when the vasculature serving the thoracic cavity constricts and blood flow is markedly reduced. Dunn and Wood sought to develop a new heart medication that would relax blood vessels and improve blood flow. In popular stories, it is often claimed that the scientists at Pfizer had no idea about the drug's side effects or its potential to improve sexual performance, but that is not entirely correct. In fact, the scientists working on this new heart drug had already found out that vasodilatation occurred throughout the body and that the drug they were working on, sildenafil citrate, might have other applications including improving blood flow to the penis, and could therefore treat erectile dysfunction.

Sildenafil was not particularly effective in combating angina, but as the researchers had suspected, it had an intriguing, prevalent, and reliable side effect in males. In fact, it was far more effective at causing erections than treating angina.

3.2 SILDENAFIL (VIAGRA)

It took many more years of clinical trials, but Viagra came to market in the United States in 1998 as the first drug to provide a relatively safe, easy, and effective way of treating erectile dysfunction. The drug rapidly became—and remains—a billion-dollar blockbuster. Competition soon arrived in the form of Cialis and Levitra. While originally marketed to older men, these drugs are often taken recreationally by younger men to enhance sexual performance, and are sometimes taken as part of a cocktail of multiple illicit drugs. Meanwhile, generic versions have come to market as the older drugs have gone off patent, and in some places these agents are available over the counter. An unforeseen benefit of the advent of these performance drugs was the initiation of a positive, responsible, and socially acceptable conversation about erectile dysfunction, which prior to the "little blue pill" had been an all-but-taboo subject. Once Viagra and similar drugs became available—and, in the United States, advertised on television—healthcare professionals and their patients felt increasingly comfortable discussing sexual dysfunction, and it soon became apparent that the problem was far more common than had been believed. With Viagra, erectile dysfunction went from being an embarrassing problem nobody talked about to a genuine medical condition that could be treated pharmacologically.

The mechanism by which sildenafil works is somewhat complex. In healthy male physiology, sexual arousal leads to increased production of cyclic guanosine monophosphate (cGMP; Fig. 3.2), which causes the muscles and tissues around the penis to relax. Sildenafil is a phosphodiesterase type 5 (PDE5) inhibitor (Fig. 3.3). In the body, cGMP is degraded naturally by the action of PDE5. A PDE5 inhibitor prevents this degradation and allows more cGMP to circulate.[1] Sildenafil was developed based on this understanding and appreciation of a nitric oxide/cGMP pathway in the body. PDE5 inhibition has also been used in drugs that treat urinary symptoms and pulmonary hypertension. The relaxation caused by PDE5 inhibition and enhanced cGMP production promotes greater blood flow, which in turn facilitates an erection. Note that these drugs do not cause an

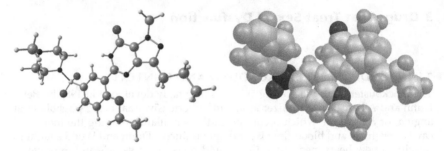

Figure 3.1 Chemical structure of sildenafil (Viagra).

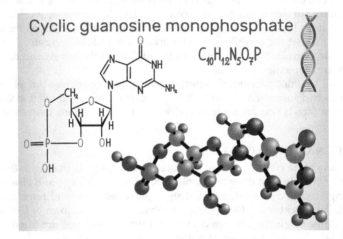

Figure 3.2 Chemical structure of cyclic guanosine monophosphate.

erection or enhance sexual desire; the erection requires sexual stimulation, while the drug merely facilitates it. Of course, sildenafil may have an indirect effect on enhancing sexual desire in that confidence about achieving and maintaining an erection might increase the desire for sexual activity.

The typical dose of sildenafil citrate is 25 or 50 mg, and the drug is ideally taken about an hour before sexual activity. However, dosing is flexible, and the drug may be taken hours before sexual activity. Doses of more than 100 mg are not recommended, and it is advised not to take sildenafil more than once a day. In fact, it should be taken at the lowest effective dose. An overdose can cause damage to the optic nerve, vomiting, diarrhea, tachycardia, and rhabdomyolysis, among other potentially life-threatening conditions. An important consideration is that PDE5 inhibitors overall have been implicated in increasing the risk of malignant melanoma, as it has been shown that PDE5 may play a role in inhibiting the proliferation of cancer cells.[2]

The most frequently reported side effects of sildenafil include headache, nasal congestion, flushing, heartburn, nausea, dizziness, and vision problems, including a bluish-green tinge to vision. The vision side effect typically resolves over time. Priapism, or a prolonged erection over four hours, is rare, but if it should it occur, it must be treated as a medical emergency. In 2007 a warning was placed on the package of Viagra that it may cause sudden loss of hearing.

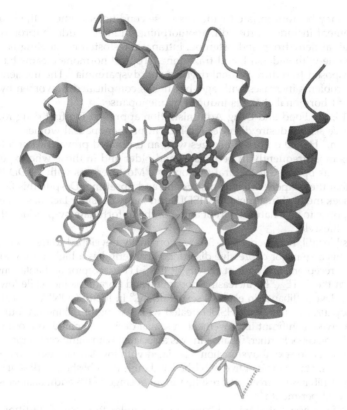

Figure 3.3 "Ribbon" model of phosphodiesterase type 5.

While enormously popular, sildenafil is not effective in about 30% of those who take it. Further, it is contraindicated in those taking nitric oxide donors, nitrates, or organic nitrites; those who have liver or kidney disease, hypotension, recent heart attack or stroke, or genetic degenerative retinal disorders; and those who have been advised to avoid sexual activity because of cardiovascular risk factors.

3.3 FLIBANSERIN (ADDYI)

Since sildenafil is not a hormonal drug and was so effective, it appeared to present a possible pharmacotherapy for sexual dysfunction in female physiology, which is less frequently and openly discussed. A small population of women treated with sildenafil reported benefits from it, but only in cases in which they had sexual desire to begin with and were troubled by insufficient arousal. For many, sexual dysfunction may be a combination of low drive and poor response, and testosterone therapy may be used to enhance sexual desire.[3]

In fact, overall, female-physiology sexual dysfunction appears less amenable to pharmacotherapy. While male-physiology sexual dysfunction mainly involves the inability to achieve or maintain an erection, female-physiology sexual dysfunction appears to be far more diverse and may include any or all of the following: low libido, poor physical genital arousal (vasocongestion, lack of lubrication), pain with intercourse, and anorgasmia. The most frequent sexual complaint is a lack of sexual desire, which the *Diagnostic and Statistical Manual of Mental Disorders, Fifth Edition 5*, calls sexual interest/arousal disorder. Drug therapy may be appropriate to address these conditions. Some of the drugs in development include alprostadil, apomorphine, bremelanotide, bupropion, intravaginal dehydroepiandrosterone, intranasal testosterone, sublingual testosterone with sildenafil, and trazodone.[4] Topical hormone creams have also been proposed to reduce vaginal dryness and dyspareunia.[3] The incidence of sexual problems increases with age, but this is complicated in women by the significant hormonal changes induced by menopause.

In 2015, the Food and Drug Administration approved the first drug for hypoactive sexual desire disorder (HSDD) in premenopausal women: flibanserin.[5] HSDD occurs at all ages with an estimated prevalence of 5%–14%.[5] It occurs most frequently in persons 40 and older and in those who experience iatrogenic as opposed to natural menopause. Most persons with HSDD (68%) report that they experience it as distressing and upsetting. Despite this fact, most do not seek medical treatment for HSDD, and many consider lack of sexual desire and response to be a natural part of aging or the unfortunate by-product of a busy modern lifestyle.

Like sildenafil, flibanserin (Fig. 3.4) was originally developed to treat something else, in this case major depressive disorder. Flibanserin has a high affinity for serotonin receptors, and agonist activity at the 5-HT_{1A} receptor and antagonist activity at the 5-HT_{2A}. By increasing dopamine and noradrenaline while lowering serotonin levels, flibanserin elevates sexual desire.[6] In the BEGONIA clinical study, premenopausal women with HSDD treated with 100 mg of flibanserin four times a day showed significant improvement in sexual desire compared to those who received a placebo. Further, flibanserin was associated with significant improvements in distress due to sexual dysfunction associated with low sexual desire compared to placebo.[7] A meta-analysis of five published and three unpublished studies ($n = 5,914$) found that flibanserin treatment resulted in an average of 0.5 additional satisfying sexual event per month.[8]

Side effects associated with flibanserin are similar to those of traditional selective serotonin reuptake inhibitors, and include dizziness, somnolence, nausea, fatigue, and dry mouth. Unlike sildenafil, which is taken as needed, flibanserin must be taken daily; while on filbanserin, women should refrain from drinking alcohol. While promising, flibanserin is relatively expensive and may not be covered by insurance.

3.4 MISCELLANEOUS

Although folklore touts the value of numerous aphrodisiac foods, such as chocolate, strawberries, asparagus, and oysters, there is no evidence other than anecdotal reports that these substances improve sexual performance or enhance

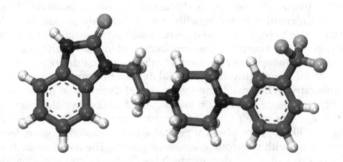

Figure 3.4 Chemical structure of flibanserin (Addyi).

sexual desire. Traditional medicines sometimes tout the sexual-enhancing powers of specific herbs such as yohimbe from West Africa and ginseng (Fig. 3.5) from East Asia. The possibility that a natural or pharmacological substance may promote sexual desire, enhance sexual performance, or improve sexual activity is not far-fetched.

Stimulants may have a stimulating effect on sexuality. For example, methylphenidate, sometimes taken as a cognitive enhancer, increases dopamine transmission and can enhance sexual arousal upon sexual stimulation. There have been anecdotal reports of enhanced sexual experiences with methylenedioxymethamphetamine (also called ecstasy), which improves serotonin transmission. It appears that dopamine, unlike serotonin, encourages

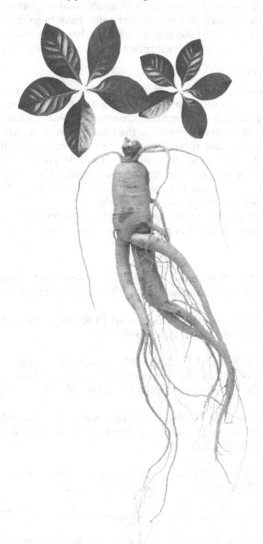

Figure 3.5 Ginseng.

sexual drive, although serotonin is associated with pleasurable emotions.[9] Methylphenidate is sometimes taken to treat attention-deficit disorder (ADD) and attention-deficit/hyperactivity disorder (ADHD). Interestingly, people with ADHD may be easily aroused sexually (due to their hyperactivity) but lack the focus to "remain in the moment" to engage in satisfying sexual activity without becoming distracted. Stimulant drugs may facilitate sexual activity in people with ADD or ADHD by promoting greater focus and calm. However, the serotonin-enhancing effects of many ADD and ADHD drugs may impair the ability to get and maintain an erection.

In other cases, stimulants may enhance sexual desire and facilitate erection but increase endurance to the point that it becomes difficult, if not impossible, to achieve orgasm. This may even result in an unpleasant or distressing experience. Illicit stimulant drugs such as methamphetamine have been implicated in risky sexual behaviors.[10,11] Likewise, cocaine use has been associated with heightened sexual desire and concomitant risky sexual behaviors.

3.5 CONCLUSION

Sildenafil represents a major breakthrough in pharmacological treatment of sexual dysfunction, but there has been less pharmacological success in treating female sexual dysfunction. Nevertheless, ongoing research shows promise, in particular low sexual interest, which is the most common sexual complaint. It appears that increased dopamine levels in the brain have a positive effect on enhancing sexual desire, while increased serotonin levels have the opposite effect, although they may be calming and pleasurable. Illicit drugs, such as methamphetamine and cocaine, may ramp up sexual desire but lead to risky sexual activity, as these drugs can impair judgment.

It is clear that we are still in the early days of serious research into the science of sexual performance and the adverse effects of drugs on it.

REFERENCES

1. Andersson KE. PDE5 inhibitors—pharmacology and clinical applications 20 years after sildenafil discovery. *Br J Pharmacol*. 2018;175(13):2554–2565.

2. Houslay MD. Melanoma, Viagra, and PDE5 inhibitors: Proliferation and metastasis. *Trends Cancer*. 2016;2(4):163–165.

3. Gregersen N, Jensen PT, Giraldi AE. Sexual dysfunction in the peri- and postmenopause. Status of incidence, pharmacological treatment and possible risks. A secondary publication. *Dan Med Bull*. 2006;53(3):349–353.

4. Belkin ZR, Krapf JM, Goldstein AT. Drugs in early clinical development for the treatment of female sexual dysfunction. *Expert Opin Investig Drugs*. 2015;24(2):159–167.

5. Gelman F, Atrio J. Flibanserin for hypoactive sexual desire disorder: Place in therapy. *Ther Adv Chronic Dis*. 2017;8(1):16–25.

6. Deeks ED. Flibanserin: First global approval. *Drugs*. 2015;75(15):1815–1822.

7. Katz M, DeRogatis LR, Ackerman R, et al. Efficacy of flibanserin in women with hypoactive sexual desire disorder: Results from the BEGONIA trial. *J Sex Med*. 2013;10(7):1807–1815.

8. Jaspers L, Feys F, Bramer WM, Franco OH, Leusink P, Laan ET. Efficacy and safety of flibanserin for the treatment of hypoactive sexual desire disorder in women: A systematic review and meta-analysis. *JAMA Intern Med.* 2016;176(4):453–462.

9. Schmid Y, Hysek CM, Preller KH, et al. Effects of methylphenidate and MDMA on appraisal of erotic stimuli and intimate relationships. *Eur Neuropsychopharmacol.* 2015;25(1):17–25.

10. Volkow ND, Wang GJ, Fowler JS, Telang F, Jayne M, Wong C. Stimulant-induced enhanced sexual desire as a potential contributing factor in HIV transmission. *Am J Psychiatry.* 2007;164(1):157–160.

11. Halkitis PN, Parsons JT, Stirratt MJ. A double epidemic: Crystal methamphetamine drug use in relation to HIV transmission among gay men. *J Homosex.* 2001;41(2):17–35.

4 Drugs That Enhance Desire or Improve Performance

4.1 INTRODUCTION

The physiology and psychology of sexual desire and sexual performance remain incompletely elucidated. Desire is likely a combination of external stimuli and internal motivations. There is likely a complex interplay between these two factors, such that an extreme outside stimulus might drive internal motivation or a strong inner drive might enhance or exaggerate otherwise inconsequential external stimuli.[1] While some typically and sometimes exclusively indulge in sexual activities in order to experience sexual pleasure—that is, responding to a hedonistic drive—for others sexuality appears to be more complex and is driven in part by hormonal cycles. Somes are more likely to be interested in sexual activity for reasons of physical pleasure, while others are more likely to be motivated by desires for relationships and intimacy; however, these objectives often overlap.[2]

4.2 DRUGS FOR FEMALE-PHYSIOLOGY AROUSAL

Although medications for male-physiology sexual dysfunction have been available for decades, female-physiology sexual dysfunction and its treatment have not received as much attention. Female-physiology sexual interest/arousal disorder (FSIAD) is defined by any three of the following symptoms: absent or greatly diminished interest in sexual activity, few or no sexual thoughts or fantasies, no initiation of or receptivity to sexual advances, decreased sexual excitement and pleasure, reduced responsiveness to sexual stimuli, and reduced or absent genital sensations during sexual activity. Since such symptoms may occur in the presence of extreme stress, traumatic junctures in relationships, mental or physical illness, licit or illicit drug use, advanced age, or some combination of these, it is important that the diagnosis exclude such possible factors. Typically, FSIAD presents clinically when women with sexual partners encounter an imbalance in their sexual needs. The absence of sexual interest or response is rarely reported by single women and almost never reported by men.

Flibanserin, described in the previous chapter, was approved by 2015 in the United States for treating low sex drive in premenopausal women. Originally developed as an antidepressant, the drug affects both serotonergic and dopaminergic neurotransmission systems. It may be used daily, but its long-term effects are not known.[2] It should not be taken with alcohol. It has been proposed that flibanserin be incorporated into combination therapy with psychological counseling to treat FSIAD.

Based on the premise that FSIAD involves both insensitivity to sexual cues and a heightened inhibitory mechanism for sexual activity, a combination of testosterone (0.5 mg sublingual) plus a phosphodiesterase type 5 inhibitor (50 mg sildenafil) has been developed.[2] A similar drug combination used testosterone plus 10 mg buspirone, a serotonin-receptor agonist. The first drug (testosterone plus sildenafil) was designed to heighten response to external sexual stimuli, while the second drug (testosterone plus buspirone) decreased inhibitions. Both drugs can be taken as needed about four hours before the effects are desired, and need not be taken daily.

Bremelanotide is a subcutaneously administered melanocortin-receptor agonist that stimulates dopamine in the brain. It may be taken on demand, with effects felt in about 45 minutes. It increases sexual desire and enhances sexual arousal.[2]

An intranasal testosterone gel promotes sexual arousal and has been favorably compared with a testosterone patch for the same use.[2] This gel increases both sexual arousal and genital response to sexual stimulation.

In addition to these drugs, other agents are currently in development, such as a synthetic peptide molecule and a memory-enhancing drug aimed at combination with cognitive behavioral therapy to help better respond neurobiologically to sexual cues and situations.[2]

4.3 AMPHETAMINES AND METHAMPHETAMINE

Amphetamines are a broad class of drug that are available in prescription and illicit forms. Methamphetamine (meth) is an illicit drug that intensifies sexual desires and may improve sexual stamina and performance.[3] "Crystal meth" is a potent, inexpensive, and potentially addictive illicit agent. As methamphetamine is the more frequently abused psychostimulant, this section will concentrate more on methamphetamine than amphetamines. Methamphetamine use disorder (MUD) is a public health crisis, and no established treatment exists for rehabilitation.[4] People with MUD exhibit cognitive dysfunctions and lower sustained attention than those who do not use meth.[5] MUD also encourages impulsivity and a desire for immediate gratification, along with deficits in executive function.[5] While cognitive deficits may improve after abstinence, some damage may be irreversible.[6]

In animal studies, methamphetamine use increases sexual interest and facilitates sexual behaviors,[7] but in self-administration studies, male rats have been shown more likely than females to use methamphetamine.[8] The effect of methamphetamine on rats with respect to increased sexual activity is dose dependent.[9]

These stimulant drugs heighten sexual arousal and stimulation, and at the same time may encourage impulsive acts and risk-taking.[10] In a study of 35 methamphetamine users in China, it was found that methamphetamine use could be significantly associated with unprotected sex and transactional sex.[11] In addition to increasing sexual desire, decreasing inhibitions, and encouraging impulsivity, methamphetamine can delay orgasm, sometimes to the point of discomfort. Many methamphetamine users may find that abnormally prolonged sexual encounters cause the sex act to lose its pleasure and appeal. Meth users are more likely than nonusers to have anonymous sex, risky sex, transactional sex, and multiple partners.

Both amphetamines and methamphetamine increase dopamine dramatically, at levels that may be too high for most brains to process on an ongoing basis. At first, this excess dopamine heightens sexual interest and responsiveness, but it also drives the body to want even more dopamine, which can be obtained from an orgasm. This sets up a vicious cycle where dopamine levels spiral out of control with consistent use. Over time, the dopamine-producing areas of the brain can become disrupted or depleted, such that the individual feels depressed, loses interest in the world, and finds little to no pleasure in sex. In some cases, dopamine depletion may cause people to initiate sexual activity but be unable to follow through, as these drugs can cause impotence and anorgasmia, particularly with prolonged use.

However, some couples use methamphetamine in their sexual relationship to facilitate or enhance coitus. In many sexual subcultures, such as in certain groups, methamphetamine use is so prevalent, it is almost expected. The behaviors associated with methamphetamine use vary by sex (see Table 4.1).[12]

Table 4.1: Sexual Behaviors that Occur in Men and Women Who Use Methamphetamine[12]

Behavior	Men	Women
Not always using condoms	90.7%	85.2%
Having multiple partners	94.2%	47.2%
Acquiring sexually transmitted infections	55.7%	56.0%
Risk factors for high-risk sexual behaviors	Using multiple stimulants. Being the client of a sex worker.	Being a migrant worker. Being employed as a sex worker.

4.4 MARIJUANA

To date, studies about the effects of marijuana on sexual behavior have produced mixed results, and the mechanisms of action by which cannabinoids may influence sexual activity remain to be elucidated. Despite current scientific interest in marijuana research, there have been relatively few studies about sex differences in cannabinoid pharmacology.[13] A recent study in female rats who were administered vaporized cannabis found an inverse dose-related response in terms of enhanced sexual behaviors, with the greatest effect at low doses.[14] A systematic review has found that marijuana may enhance the subjective experience of sex , but at high doses it might contribute to erectile dysfunction.[15] In a survey of women who used marijuana (n = 373), 34% said they used it prior to sex, and those women had 2.13 times higher odds of achieving a satisfactory orgasm compared to those who did not use marijuana before sex. In general, women who used marijuana frequently had 2.10 times higher odds of satisfactory orgasms than infrequent users.[16]

In a survey of 216 people who reported having sex while under the influence of marijuana, 39% said marijuana improved their sexual experience and 25% said it sometimes did, while 5% said it made sex less pleasurable. However, 74% said marijuana increased their tactile sensitivity, and 66% said it made their orgasms more intense. Notably, 70% of respondents said that marijuana aided them in relaxing during sex.[17] In a survey of 28,176 women (mean age, 29.9 years) and 22,943 men (mean age, 29.5 years), there was a positive association between sexual frequency and marijuana use in both sexes across all demographic groups.[18] There is preliminary evidence suggesting that marijuana use is associated with hypersexuality, which is defined as sexual behaviors and desires that persist despite a desire to control or decrease them.[19]

4.5 3,4-METHYLENEDIOXYMETHAMPHETAMINE (MDMA)

MDMA, sometimes called ecstasy, is a drug that may be prescribed in certain psychotherapeutic settings but is more frequently taken illicitly for recreational purposes. MDMA is widely known as a "club drug," and its use carries less social stigma than meth or heroin. MDMA usually comes in pill form; a powder variant, known as Molly, often contains potentially dangerous adulterants, such as methamphetamine or other drugs. In a study of 24 healthy subjects, MDMA was shown to produce feelings of happiness, trust, and well-being, plus it reduced anxiety. However, in the Facial Emotion Recognition Task (FERT), MDMA led to frequent misclassifications of emotion—users tended to see more happy and more positive emotions.[20]

MDMA increases the body's production of cortisol, prolactin, and oxytocin,[20] and enhances serotonin neurotransmission.[21] This is interesting, because drugs that stimulate serotonin production generally impair sexual arousal and diminish sexual performance, whereas those that affect the dopaminergic system do the opposite—that is, they increase sexual drive. However, MDMA enhances empathy, sociability, and feelings of well-being, and may cause an individual to feel closer to others even though it does not appear to have a direct stimulating effect on sex drive.[21] Reports of MDMA stimulating sexual desire or improving sexual pleasure or responsiveness remain inconsistent.[21] It may differentially affect the desire for the act, and feelings of sensuality and emotional union.[21] That is, MDMA may increase the desire to interact with others in such a way that can lead to sexual activity, although it does not seem to increase interest in sexual activity *per se*.[21]

4.6 METHYLPHENIDATE

Methylphenidate (better known by its trade name Ritalin) is a drug developed to treat attention-deficit/hyperactivity disorder but is sometimes taken as a stimulant or to temporarily enhance cognitive function or focus. It is also sometimes taken for recreational purposes. It increases dopamine and norepinephrine neurotransmission by way of inhibiting their reuptake. (Thus, it functions inversely to MDMA, which causes the release of serotonin.) When MDMA and methylphenidate were compared in adults viewing erotic images, methylphenidate induced greater sexual interest and stimulation than MDMA, and participants said that it increased their level of sexual arousal more than MDMA.[21]

The combined use of MDMA and methylphenidate does not increase their psychoactive or sex-enhancing properties, but it does increase cardiovascular risk and adverse events.[22]

4.7 COCAINE

Cocaine enjoys a curious reputation among many users as an aphrodisiac, although it is as likely to impair sexual performance as enhance it. On the one hand, it increases sexual desire in a dose-dependent fashion and may also encourage risk-taking behaviors and impulsivity.[23] As a psychomotor stimulant, cocaine may pique sexual interest, heighten arousal, improve performance, and enhance sexual pleasure, particularly with occasional as opposed to habitual use. With long-term use, it may contribute to erectile dysfunction or anorgasmia.[24] Cocaine stimulates the dopaminergic system (which encourages sexual activity), but if this system becomes depleted or deranged with persistent cocaine use, it may inhibit sexual arousal, decrease libido, and impair sexual response and performance. In fact, in one study, sexual dysfunction occurred in 62% of regular cocaine users.[25]

It has been suggested that the role of cocaine as a drug to enhance sexual encounters may have more to do with the context of its use than its direct pharmacological properties. Cocaine users who indulge in drug use with a partner may feel a special sort of celebratory bonding over sharing an illicit and expensive substance and participating in a forbidden activity. Cocaine also works to diminish social inhibitions, heighten impulsivity, and increase risk-taking behaviors. While it often diminishes sexual performance, it is more frequently used with intimate partners than other drugs, such as heroin.[26] Because cocaine enjoys a reputation among certain illicit drug users as a "sexy" drug, its alleged aphrodisiac benefits may be more of a psychological effect rather than an actual pharmacologically induced one.[26]

4.8 OPIOIDS

Opioids, such as morphine and oxycodone, inhibit libido. Sexual dysfunction is common in people who take opioids long-term; heroin-dependent individuals have 34%–85% rates of sexual dysfunction. Opioid-associated sexual adverse events include erectile dysfunction, premature ejaculation, retarded ejaculation, and diminished libido in men, and dyspareunia and vaginal dryness.[27] Even people taking prescription opioids report sexual dysfunction; 82% of those taking acute opioid therapy and 69% of those on chronic opioid therapy report dissatisfaction with their sex lives.[28]

Over time, opioids can reduce the levels of circulating gonadotropin-releasing hormone, which originates from the hypothalamus; this causes the pituitary gland to secrete less luteinizing hormone and follicle-stimulating hormone, which in turn decreases testosterone production in the testis. This condition, known as hypogonadism, has been associated with sexual dysfunction, adiposity, muscle wasting syndrome, infertility, and osteoporosis.[29] Hypogonadism may be treated with testosterone, but few, even those known to regularly take prescription or illicit opioids, are ever screened for the condition. Opioid use may result in amenorrhea.[30]

4.9 SUBSTANCE-USE DISORDERS

Despite the fact that the use of illicit substances may alter sexual behaviors, there is very little study into this topic. In a survey of 180 people being treated for substance-use disorders (alcohol, meth, cocaine, opioids, and others), all respondents said that their preferred substance changed their thoughts about sex and sexual behaviors. More than half said that their substance of choice enhanced their sexual pleasure, although in some cases they said it led to risky sexual behaviors.[31] Studies suggest that the odds of risky sexual behaviors increase when one or both partners use illicit drugs or alcohol.[32]

Polydrug abuse involves the misuse of various illicit and licit drugs, often taken opportunistically. Polydrug abusers often take opioids, cocaine, marijuana, and benzodiazepines on a fairly regular basis and add other agents, such as MDMA, as available. Polydrug abuse is associated with erectile dysfunction.[24] Paraphilic sexual desires and behaviors were found in a study in Brazil to occur in about half of people with some substance-use disorder, including polydrug abusers.[24]

4.10 NATURAL SUBSTANCES

Despite nutraceuticals that promise to enhance sexual activity, there has been little study to offer evidence that natural substances can serve as aphrodisiacs. This is not to say that some of these substances are ineffective; they have just not been thoroughly studied. Furthermore, it is possible that some natural products that may increase blood flow, stabilize hormonal fluctuations, or improve overall well-being may have a positive effect on sexual desire and performance as well. Natural products that enhance a sense of well-being may likewise have a positive, if indirect, effect on sex. Sexual activity is complex, and many factors work together in terms of sexual desire, sexual response, and orgasm.

Foods claimed to be aphrodisiacs include chocolate, honey, hot chilies, strawberries, papaya, oysters, avocados, watermelon, or herbs such as fenugreek, ginseng, and ginkgo biloba. Maca, the South American root vegetable *Lepidium meyenii*, is ground to a powder to be taken for sexual benefits, and there is some evidence that it improves libido in postmenopausal women and may counteract erectile dysfunction.[33]

Ylang-ylang, with its volatile constituent β-caryophyllene, is thought to reduce sexual anxiety and is traditionally placed on the wedding bed in Indonesia. In a study in which 19 women in the follicular phase of the menstrual cycle were exposed to aromatherapy with β-caryophyllene, testosterone levels as measured in saliva increased, but not estrogen, which suggests that ylang-ylang may promote sexual desire.[34]

Ginseng is considered a functional food and an adaptogenic herb, with purported claims that it is a potent antioxidant, anti-inflammatory, and neuroprotective drug that fights fatigue and enhances sex drive.[35] Known as the "king of all herbs," ginseng has been used for centuries in East Asia as a tonic for improved vitality and virility.[36] It is one of the few natural substances that have been rigorously studied in terms of sexual symptoms. A double-blind placebo-controlled study of 45 men with moderate to severe erectile dysfunction reported improvement over eight weeks after taking three doses a day of 900 mg of Korean red ginseng.[37] In a 12-week study, 60 men showed significant improvement in the ability to achieve and maintain an erection after taking 1000 mg of Korean red ginseng three times a day.[38] Ginseng and its ginsenoside active constituents induce nitric oxide synthesis in endothelial cells; it has been hypothesized that this action relaxes the perivascular nerves in the smooth muscles, improving blood flow to the penis.[39] There has been less investigation into the effects of ginseng on sexual behaviors in women, and in one trial with 23 premenopausal women, an eight-week course of Korean red ginseng was not superior to placebo in improving sex drive and sexual pleasure.[40]

REFERENCES

1. Rudzinskas SA, Williams KM, Mong JA, Holder MK. Sex, drugs, and the medial amygdala: A model of enhanced sexual motivation in the female rat. *Front Behav Neurosc.* 2019;13:203–203.

2. Both S. Recent developments in psychopharmaceutical approaches to treating female sexual interest and arousal disorder. *Curr Sex Health Rep.* 2017;9(4):192–199.

3. Rawson RA, Washton A, Domier CP, Reiber C. Drugs and sexual effects: Role of drug type and gender. *J Subst Abuse Treat.* 2002;22(2):103–108.

4. Stauffer CS, Moschetto JM, McKernan S, et al. Oxytocin-enhanced group therapy for methamphetamine use disorder: Randomized controlled trial. *J Subst Abuse Treat.* 2020;116:108059.

5. Bernhardt N, Petzold J, Groß C, et al. Neurocognitive dysfunctions and their therapeutic modulation in patients with methamphetamine dependence: A pilot study. *Front Psychiatry.* 2020;11:581.

6. Farhadian M, Akbarfahimi M, Hassani Abharian P, Hosseini SG, Shokri S. Assessment of executive functions in methamphetamine-addicted individuals: Emphasis on duration of addiction and abstinence. *Basic Clin Neurosci.* 2017;8(2):147–153.

7. Holder MK, Mong JA. Methamphetamine enhances paced mating behaviors and neuroplasticity in the medial amygdala of female rats. *Horm Behav.* 2010;58(3):519–525.

8. Daiwile AP, Jayanthi S, Ladenheim B, et al. Sex differences in escalated methamphetamine self-administration and altered gene expression associated with incubation of methamphetamine seeking. *Int J Neuropsychopharmacol.* 2019;22(11):710–723.

9. Frohmader KS, Bateman KL, Lehman MN, Coolen LM. Effects of methamphetamine on sexual performance and compulsive sex behavior in male rats. *Psychopharmacology.* 2010;212(1):93–104.

10. Ahuja N, Schmidt M, Dillon PJ, Alexander AC, Kedia S. Online narratives of methamphetamine use and risky sexual behavior: Can shame-free guilt aid in recovery? *Arch Sex Behav.* 2020.

11. Liu L, Chai X. Pleasure and risk: A qualitative study of sexual behaviors among Chinese methamphetamine users. *J Sex Res.* 2020;57(1):119–128.

12. Saw YM, Saw TN, Chan N, Cho SM, Jimba M. Gender-specific differences in high-risk sexual behaviors among methamphetamine users in Myanmar-China border city, Muse, Myanmar: Who is at risk? *BMC Public Health.* 2018;18(1):209.

13. Craft RM, Marusich JA, Wiley JL. Sex differences in cannabinoid pharmacology: A reflection of differences in the endocannabinoid system? *Life Sci.* 2013;92(8-9):476–481.

14. Mondino A, Fernández S, Garcia-Carnelli C, et al. Vaporized cannabis differentially modulates sexual behavior of female rats according to the dose. *Pharmacol Biochem Behav.* 2019;187:172814.

15. Rajanahally S, Raheem O, Rogers M, et al. The relationship between cannabis and male infertility, sexual health, and neoplasm: A systematic review. *Andrology.* 2019;7(2):139–147.

16. Lynn BK, López JD, Miller C, Thompson J, Campian EC. The relationship between marijuana use prior to sex and sexual function in women. *Sexual Med.* 2019;7(2):192–197.

17. Wiebe E, Just A. How cannabis alters sexual experience: A survey of men and women. *J Sex Med.* 2019;16(11):1758–1762.

18. Sun AJ, Eisenberg ML. Association between marijuana use and sexual frequency in the united states: A population-based study. *J Sex Med.* 2017;14(11):1342–1347.

19. Slavin MN, Kraus SW, Ecker A, et al. Marijuana use, marijuana expectancies, and hypersexuality among college students. *Sex Addict Compulsivity.* 2017;24(4):248–256.

20. Dolder PC, Müller F, Schmid Y, Borgwardt SJ, Liechti ME. Direct comparison of the acute subjective, emotional, autonomic, and endocrine effects of MDMA, methylphenidate, and modafinil in healthy subjects. *Psychopharmacology*. 2018;235(2):467–479.

21. Schmid Y, Hysek CM, Preller KH, et al. Effects of methylphenidate and MDMA on appraisal of erotic stimuli and intimate relationships. *Eur Neuropsychopharmacol*. 2015;25(1):17–25.

22. Hysek CM, Simmler LD, Schillinger N, et al. Pharmacokinetic and pharmacodynamic effects of methylphenidate and MDMA administered alone or in combination. *Int J Neuropsychopharmacol*. 2014;17(3):371–381.

23. Johnson MW, Herrmann ES, Sweeney MM, LeComte RS, Johnson PS. Cocaine administration dose-dependently increases sexual desire and decreases condom use likelihood: The role of delay and probability discounting in connecting cocaine with HIV. *Psychopharmacology*. 2017;234(4):599–612.

24. Clemente J, Diehl A, Santana P, da Silva CJ, Pillon SC, Mari JJ. Erectile dysfunction symptoms in polydrug dependents seeking treatment. *Subst Use Misuse*. 2017;52(12):1565–1574.

25. Cocores JA, Miller NS, Pottash AC, Gold MS. Sexual dysfunction in abusers of cocaine and alcohol. *Am J Drug Alcohol Abuse*. 1988;14(2):169–173.

26. Kopetz CE, Reynolds EK, Hart CL, Kruglanski AW, Lejuez CW. Social context and perceived effects of drugs on sexual behavior among individuals who use both heroin and cocaine. *Exp Clin Psychopharmacol*. 2010;18(3):214–220.

27. Grover S, Mattoo SK, Pendharkar S, Kandappan V. Sexual dysfunction in patients with alcohol and opioid dependence. *Indian J Psychol Med*. 2014;36(4):355–365.

28. Rapaport L. Opioids tied to bad sex life and lack of desire. *Physician's Weekly*. http://physiciansweekly.com/opioids-tied-to-bad/. Published 2018. Accessed August 25, 2020.

29. Baillargeon J, Raji MA, Urban RJ, et al. Opioid-induced hypogonadism in the United States. *Mayo Clin Proc Innov Qual Outcomes*. 2019;3(3):276–284.

30. Reddy RG, Aung T, Karavitaki N, Wass JAH. Opioid induced hypogonadism. *BMJ Clin Res Ed*. 2010;341:c4462–c4462.

31. Bosma-Bleeker MH, Blaauw E. Substance use disorders and sexual behavior: The effects of alcohol and drugs on patients' sexual thoughts, feelings and behavior. *Addict Behav*. 2018;87:231–237.

32. Brown RE, Turner C, Hern J, Santos GM. Partner-level substance use associated with increased sexual risk behaviors among men who have sex with men in San Francisco, CA. *Drug Alcohol Depend*. 2017;176:176–180.

33. West E, Krychman M. Natural aphrodisiacs—A review of selected sexual enhancers. *Sex Med Rev.* 2015;3(4):279–288.

34. Tarumi W, Shinohara K. Olfactory exposure to β-caryophyllene increases testosterone levels in women's saliva. *Sex Med.* 2020.

35. Patel S, Rauf A. Adaptogenic herb ginseng (Panax) as medical food: Status quo and future prospects. *Biomed Pharmacother.* 2017;85:120–127.

36. Carota G, Raffaele M, Sorrenti V, Salerno L, Pittalà V, Intagliata S. Ginseng and heme oxygenase-1: The link between an old herb and a new protective system. *Fitoterapia.* 2019;139:104370.

37. Hong B, Ji YH, Hong JH, Nam KY, Ahn TY. A double-blind crossover study evaluating the efficacy of Korean red ginseng in patients with erectile dysfunction: A preliminary report. *J Urol.* 2002;168(5):2070–2073.

38. de Andrade E, de Mesquita AA, Claro Jde A, et al. Study of the efficacy of Korean Red Ginseng in the treatment of erectile dysfunction. *Asian J Androl.* 2007;9(2):241–244.

39. Leung KW, Wong AS. Ginseng and male reproductive function. *Spermatogenesis.* 2013;3(3):e26391–e26391.

40. Chung HS, Hwang I, Oh KJ, Lee MN, Park K. The effect of Korean Red Ginseng on sexual function in premenopausal women: Placebo-controlled, double-blind, crossover clinical trial. *Evid Based Complement Alternat Med.* 2015;2015:913158–913158.

5 Male-Physiology Fertility

5.1 INTRODUCTION

Industrialized Western nations have been observing a slow but progressive decline in male fertility in the past half century. Investigators have observed an overall decline in testosterone levels and sperm quality, which have been attributed to lifestyle factors, substance abuse, and environmental issues.[1] It is estimated that about 13% of heterosexual couples are infertile—that is, the couple is unable to get pregnant. In about 50% of these cases, male factor infertility (MFI) is the culprit, typically in terms of sperm production or sperm delivery.[2] Both testosterone and sperm are made in the testicles. While testosterone circulates in the body, sperm is stored in the testicles. During sexual arousal, sperm travels through the epididymis, a tube behind each testicle. Just prior to ejaculation, sperm moves out of the epididymis into the vas deferens, which connects to an ejaculatory duct in the seminal vesicle (Fig. 5.1). Upon ejaculation, the sperm (Fig. 5.2) is mixed with fluids formed by the prostate and the seminal vesicles, creating semen. Semen is then ejaculated through the urethra and out of the penis. If the man is having vaginal sex with a woman, the semen then enters the vagina and travels upward through the cervix into the uterus and from there to the fallopian tubes. If a sperm is able to navigate this journey and meet up with an egg (which is in proper position to be fertilized only a few days a month), the egg is then fertilized (Fig. 5.3)—in other words, conception occurs (Fig. 5.4).

There are any number of potential problems that can prevent successful fertilization, of which the most common is inadequate sperm production. Oligospermia occurs when the man makes too little sperm; azoospermia occurs when no sperm are created at all. Sperm may also be oddly shaped, poorly motile, or immature. Chromosomal defects of sperm may render them infertile; a sperm cell carries half of the DNA to the egg, and defective DNA can result in infertility. Oligospermia and azoospermia are subdivided into pretesticular, testicular, and posttesticular, depending on where the impairment occurs. For example, pretesticular azoospermia is caused by a defect in the pituitary gland, which fails to stimulate the testes to properly produce sperm. This is usually the result of an endocrine imbalance.[2] Testicular azoospermia occurs because of testicular dysfunction, and posttesticular azoospermia is typically caused by some obstruction that prevents or hinders proper ejaculation. While some problems with sperm may be genetic in origin, lifestyle choices, including medications, can cause sperm problems.

5.2 MEDICATIONS THAT CAN REDUCE SPERM PRODUCTION

About half of all problems with male fertility are caused by abnormal sperm production.[3] Many commonly prescribed medications can suppress sperm production, such as drugs that treat arthritis, depression, and digestive disorders, as well as antibiotics, antihypertensives, and cancer drugs. Illicit or recreational drugs, such as anabolic steroids, cocaine, and marijuana, can also reduce sperm count. Although not considered a pharmacological agent, alcohol has a similar inhibitory effect on sperm production.

Despite the fact that male infertility is a prevalent and distressing problem, there is a paucity of studies in terms of how specific drugs affect sperm parameters. Some drugs, such as sildenafil, may increase sperm motility and improve sperm

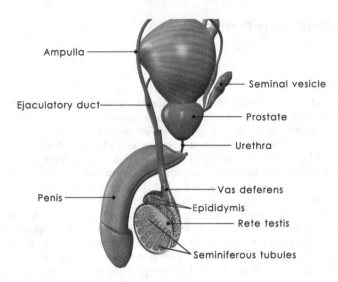

Ampulla

Seminal vesicle

Ejaculatory duct

Prostate

Urethra

Vas deferens

Penis

Epididymis

Rete testis

Seminiferous tubules

Figure 5.1 Anatomical location of ejaculatory duct.

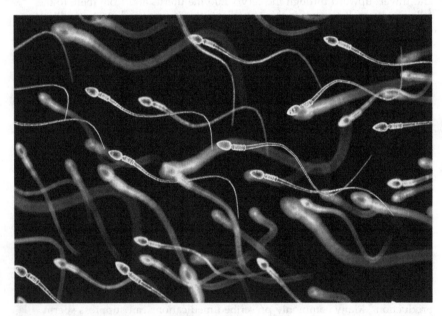

Figure 5.2 Spermatozoa (sperm).

parameters.[4] Much of what is known about the effect of commonly prescribed drugs on male fertility comes from small studies and case reports.

5.2.1 Alpha-Adrenergic Blockers ('Alpha Blockers')

Alpha-adrenergic blockers are indicated for the treatment of lower urinary tract infection associated with benign prostatic hyperplasia. These agents have a high binding affinity for both alpha-adrenergic receptors and dopamine and serotonin

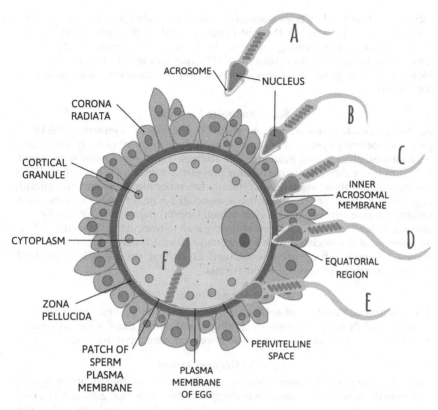

Figure 5.3 Fertilization of egg (ovum, oocyte).

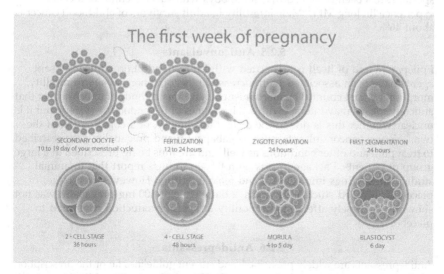

Figure 5.4 First week of pregnancy.

receptors. Their use has been associated with retrograde ejaculation and even failure to ejaculate, which is thought to be the result of their effects on the dopaminergic and serotonergic systems. In a study of men treated with tamsulosin, motile sperm decreased by 14% from baseline in five days.[5] However, not all alpha-adrenergic blockers have the same effects on sperm parameters.[6]

5.2.2 Anabolic Steroids

Anabolic steroids define a group of about 30 drugs used to enhance athletic performance and physical appearance that are frequently taken by bodybuilders, professional athletes, and even amateur athletes. Their use is more widespread than commonly thought, with a lifetime prevalence rate of 6.4% for men and 1.6% for women.[7] Anabolic steroids are typically taken nonmedically and are officially prohibited in many sports. Their use is associated with a number of potentially serious adverse effects, including on semen (sperm) parameters.[1] Anabolic steroids reduce testosterone levels and may lead to testicular dysfunction.[1] Their long-term use is associated with hypogonadism, which may be gradually reversed when the drugs are discontinued. Over time, the use of high levels of anabolic steroids can result in azoospermia.[1]

5.2.3 Antibiotics

There is limited evidence of antibiotics affecting semen quality, except for the drug class of fluoroquinolones—including brand names like Cipro, Levaquin, and Avelox—which may have adverse effects on semen parameters.

5.2.4 Anticancer Drugs

Testicular tissue is very susceptible to damage from chemotherapeutic drugs and radiation therapy. Certain anticancer drugs may be considered gonadotoxic. With more and more people surviving cancer or living with managed cancer, this may be an important consideration before embarking on therapy. For example, a person with cancer may wish to take steps before cancer treatment to preserve sperm. It has been observed that some boys who survive childhood cancer experience lifelong MFI.[8] Male infertility in adult survivors of childhood cancer is about 46%.[8]

5.2.5 Anticonvulsants

Epilepsy in and of itself is associated with MFI.[9] In addition, anticonvulsant agents have been associated with increased rates of abnormal sperm motility, morphology, and count. The mechanism behind this is thought to be the fact that anticonvulsants may interact with sex hormones of the endocrine system, but it is unclear whether this is due to the drug, the epilepsy, or a combination of the two.[3] Newer anticonvulsants, such as gabapentin and pregabalin, are prescribed to treat pain and other conditions as well, and are thus being prescribed in a large group of patients. Package inserts with these products report that in animal studies, these drugs impair male and female fertility. However, in a large, placebo-controlled study in humans, a daily dose of 600 mg pregabalin was not shown to adversely affect sperm motility or sperm production compared to placebo.[10]

5.2.6 Antidepressants

Antidepressants are widely prescribed, and current guidelines favor the prescription of selective serotonin reuptake inhibitors (SSRIs), which may cause MFI. However, it cannot be known to what extent an SSRI reduces sperm production, because depression itself may lower testosterone levels, diminish sexual drive, and possibly contribute to MFI apart from pharmacological therapy. A case report found that two people who took SSRIs for depression had medication-associated alterations in sperm motility and sperm concentration that improved at two months after the SSRIs were discontinued.[11] Data from three large prospective cohort studies in 100 people show reduced sperm concentration, impaired mobility, and higher levels of abnormal sperm cells and DNA fragmentation in people who took an SSRI for three months or longer.[12] The mechanism behind this is suspected to be the endocrine disruption, as evidenced by the fact that children taking SSRIs exhibit delays in growth compared to their peers who do not take SSRIs.[12]

Dapoxetine is an SSRI currently marketed to treat premature ejaculation. While dapoxetine is indicated for this use, SSRIs were previously already used off-label to treat this condition. The drug was shown in a randomized double-blind placebo-controlled trial to delay intravaginal ejaculatory latency time significantly compared to placebo.[13]

SSRIs have been demonstrated to have a spermicidal effect *in vitro*.[14] However, there remains a lack of evidence from large studies to show that SSRIs may reduce male fertility.

5.2.7 Antihistamines

The long-term use of antihistamines or other allergy medications may adversely affect male fertility with long-term consequences. Since histamines play a crucial role in erection and ejaculation, their long-term inhibition with allergy medications may have adverse effects on fertility.

5.2.8 Antiretrovirals

Highly active antiretroviral therapy (HAART) is prescribed to people with HIV, and lowers mortality but increases the rates of certain serious side effects, such as peripheral lipodystrophy, insulin resistance, and hyperlipidemia.[15] It may also adversely affect fertility. HAART is a combination therapy, and it has been shown that saquinavir (a drug often used in this treatment) decreases sperm motility *in vitro*.[16] In a study of 34 men at an HIV outpatient clinic in the Netherlands, sperm motility decreased significantly over the 48-week course of the study, although other semen parameters remained in the normal range.[17] Since HAART is an important and lifesaving therapy, it is not appropriate to consider its discontinuation even in men who wish to father children.

5.2.9 Aspirin

The literature contains a variety of sources suggesting that the use of aspirin may adversely affect semen quality, but there are no definitive studies to support this hypothesis. It is thought that aspirin decreases testosterone synthesis, lowers the production of testicular prostaglandins, and reduces the production of nitric oxide in the semen, and in that way may enhance oxidative injury to sperm cells.[18]

5.2.10 Calcium-Channel Blockers

Calcium-channel blockers are often prescribed to treat hypertension and certain cardiac conditions. Their mechanism of action inhibits the flow of free calcium ions through calcium channels and into cells. Calcium ions affect sperm in different ways depending on the development of the sperm. In sperm cells that have not fully matured, free calcium ions appear to increase motility, but some data have emerged to suggest that taking calcium-channel blockers reduces fertility and thus must exert a negative effect on mature sperm.[19]

5.2.11 Marijuana

The endocannabinoid system of the body affects reproduction. Long-term marijuana use reduces testosterone levels and causes abnormal semen in a dose-dependent fashion.[20] In a study of 1,215 men, the use of marijuana more than once a week was associated with lower sperm concentrations and lower sperm counts, a condition which was worsened when other licit or illicit drugs were also used.[21]

5.2.12 Testosterone Replacement Therapy

Taking exogenous testosterone may result in a sudden increased level of circulating testosterone temporarily, but markedly suppressed sperm production. This effect is thought to be temporary.

5.2.13 Other Drugs

The odds ratio of oligospermia occurring with cocaine use is 2.1.[22] In animal studies, methamphetamine increases cellular apoptosis in testicular germ cells.[23] Opioids reduce testosterone production, but it has been theorized that some do this to a greater degree than others; buprenorphine, for example, has less effect on testosterone production than methadone.[24] This hypothesis has been challenged, and it may be that all opioids decrease testosterone levels to more or less the same extent.[3]

5.3 CONCLUSION

Male infertility is increasing in developed nations for a variety of reasons, largely attributed to a more toxic environment, poor lifestyle choices, and increased rates of substance abuse. In addition, parenthood is often delayed , so that age-related problems associated with fertility may also emerge. While these factors play a role in infertility, some fertility problems are related to prescribed and illicit drugs. This association is not thoroughly studied, and in many cases there are only a few reports in the literature or hypotheses to guide us. Many drugs can adversely affect sperm production, sperm motility, and other semen parameters.

REFERENCES

1. Duca Y, Aversa A, Condorelli RA, Calogero AE, La Vignera S. Substance abuse and male hypogonadism. *J Clin Med*. 2019;8(5).

2. Singh K, Jaiswal D. Human male infertility: A complex multifactorial phenotype. *Reprod Sci*. 2011;18(5):418–425.

3. Brezina PR, Yunus FN, Zhao Y. Effects of pharmaceutical medications on male fertility. *J Reprod Infertil*. 2012;13(1):3–11.

4. Pomara G, Morelli G, Canale D, et al. Alterations in sperm motility after acute oral administration of sildenafil or tadalafil in young, infertile men. *Fertil Steril*. 2007;88(4):860–865.

5. Hellstrom WJ, Sikka SC. Effects of alfuzosin and tamsulosin on sperm parameters in healthy men: Results of a short-term, randomized, double-blind, placebo-controlled, crossover study. *J Androl*. 2009;30(4):469–474.

6. Andersson KE, Wyllie MG. Ejaculatory dysfunction: Why all alpha-blockers are not equal. *BJU Int*. 2003;92(9):876–877.

7. Sagoe D, Molde H, Andreassen CS, Torsheim T, Pallesen S. The global epidemiology of anabolic-androgenic steroid use: A meta-analysis and meta-regression analysis. *Ann Epidemiol.* 2014;24(5):383–398.

8. Delessard M, Saulnier J, Rives A, Dumont L, Rondanino C, Rives N. Exposure to chemotherapy during childhood or adulthood and consequences on spermatogenesis and male fertility. *Int J Mol Sci.* 2020;21(4).

9. Webber MP, Hauser WA, Ottman R, Annegers JF. Fertility in persons with epilepsy: 1935–1974. *Epilepsia.* 1986;27(6):746–752.

10. Sikka SC, Chen C, Almas M, Dula E, Knapp LE, Hellstrom WJ. Pregabalin does not affect sperm production in healthy volunteers: A randomized, double-blind, placebo-controlled, noninferiority study. *Pain Pract.* 2015;15(2):150–158.

11. Tanrikut C, Schlegel PN. Antidepressant-associated changes in semen parameters. *Urology.* 2007;69(1):185.e185–187.

12. Semen abnormalities with SSRI antidepressants. *Prescrire Int.* 2015;24(156): 16–17.

13. McMahon CG. Dapoxetine: A new option in the medical management of premature ejaculation. *Ther Adv Urol.* 2012;4(5):233–251.

14. Kumar VS, Sharma VL, Tiwari P, et al. The spermicidal and antitrichomonas activities of SSRI antidepressants. *Bioorg Med Chem Lett.* 2006;16(9):2509–2512.

15. Carr A, Samaras K, Burton S, et al. A syndrome of peripheral lipodystrophy, hyperlipidaemia and insulin resistance in patients receiving HIV protease inhibitors. *Aids.* 1998;12(7):F51–58.

16. Ahmad G, Moinard N, Jouanolou V, Daudin M, Gandia P, Bujan L. In vitro assessment of the adverse effects of antiretroviral drugs on the human male gamete. *Toxicol In Vitro.* 2011;25(2):485–491.

17. van Leeuwen E, Wit FW, Repping S, et al. Effects of antiretroviral therapy on semen quality. *Aids.* 2008;22(5):637–642.

18. Banihani SA. Effect of aspirin on semen quality: A review. *Andrologia.* 2020;52(1):e13487.

19. Katsoff D, Check JH. A challenge to the concept that the use of calcium channel blockers causes reversible male infertility. *Hum Reprod.* 1997;12(7):1480–1482.

20. Kolodny RC, Masters WH, Kolodner RM, Toro G. Depression of plasma testosterone levels after chronic intensive marihuana use. *N Engl J Med.* 1974;290(16):872–874.

21. Gundersen TD, Jørgensen N, Andersson AM, et al. Association between use of marijuana and male reproductive hormones and semen quality: A study among 1,215 healthy young men. *Am J Epidemiol.* 2015;182(6):473–481.

22. Bracken MB, Eskenazi B, Sachse K, McSharry JE, Hellenbrand K, Leo-Summers L. Association of cocaine use with sperm concentration, motility, and morphology. *Fertil Steril*. 1990;53(2):315–322.

23. Yamamoto Y, Yamamoto K, Hayase T, Abiru H, Shiota K, Mori C. Methamphetamine induces apoptosis in seminiferous tubules in male mice testis. *Toxicol Appl Pharmacol*. 2002;178(3):155–160.

24. Bliesener N, Albrecht S, Schwager A, Weckbecker K, Lichtermann D, Klingmüller D. Plasma testosterone and sexual function in men receiving buprenorphine maintenance for opioid dependence. *J Clin Endocrinol Metab*. 2005;90(1):203–206.

6 Female-Physiology Fertility

6.1 INTRODUCTION

Fertility requires ovulation, the monthly release of an egg from a follicle in at least one ovary into the fallopian tubes, which connect the ovaries to the uterus (Fig. 6.1). If the egg encounters viable sperm at this point, it is fertilized and conception occurs. The fertilized egg then travels through the fallopian tube and implants itself into the endometrium lining the uterus. The monthly release of an egg is governed by hormonal cycles (Fig. 6.2), and two hormones in particular affect the development of the egg in the ovary: follicle-stimulating hormone (FSH) and luteinizing hormone (LH). If the egg is not fertilized, it is flushed from the body along with the endometrial lining of the uterus in the monthly menstrual flow (Fig. 6.3). Certain criteria must be met for fertility: a viable egg must be produced, the fallopian tubes must be clear and functional, and once fertilized, the egg must be able to implant itself in the uterus. Any shortfall in these three conditions can result in infertility.

Ovulation occurs at the midpoint of a menstrual cycle (Fig. 6.4), for example, day 14 of a 28-day cycle. However, the exact time of fertility may span several days around this time; in addition, sperm can survive up to three days in vivo, so there are likely several days each month when a pregnancy can be initiat.

It is important to distinguish between fertility (the ability to conceive) and fecundity (the ability to bring a child to term). About 6% of women between the ages of 15 and 44 are infertile after one year of trying to get pregnant, and 12% of all women have impaired fecundity.[1] Fertility issues affect both men and women, but fecundity is a female-only issue

Many things influence fertility and fecundity, the most important of which by far is age. A woman is born with all of the eggs she will ever have, but the number of viable eggs decreases markedly with age. This in part explains age-related fertility issues and the "biological clock" (Fig. 6.5). For example, at age 36, a woman is half as likely to conceive as at age 20. Women over 40 have a relatively low chance of conceiving naturally, although in vitro fertilization and other techniques can help older women conceive and bear children.

However, other factors may prevent pregnancy. Hormonal disorders, such as polycystic ovary syndrome (PCOS) and hyperprolactinemia, can cause infertility. Thyroid problems, both hyperthyroidism and hypothyroidism, can do the same. Fibroid tumors in the uterus or cervix, damage to the fallopian tubes, or pelvic scarring or adhesions can also make conception difficult or impossible. Pelvic inflammatory disease and endometriosis can result in infertility. Early menopause, defined as primary ovarian insufficiency before age 40, may occur for natural or iatrogenic reasons. Other factors that may affect fertility include obesity, sexually transmitted diseases, and unhealthy lifestyle. Exposure to certain toxins, such as dangerous pesticides, may impair fertility. Extensive and grueling exercise, such as occurs with professional or Olympic athletes, can reduce progesterone production and inhibit ovulation, resulting in reversible infertility.

6.2 DRUGS USED FOR INFERTILITY

There is a paucity of research about drugs and infertility, but it is evident that certain substances may adversely affect an ability to conceive. Fertility rests on the production of a viable egg, which involves complex bidirectional communications

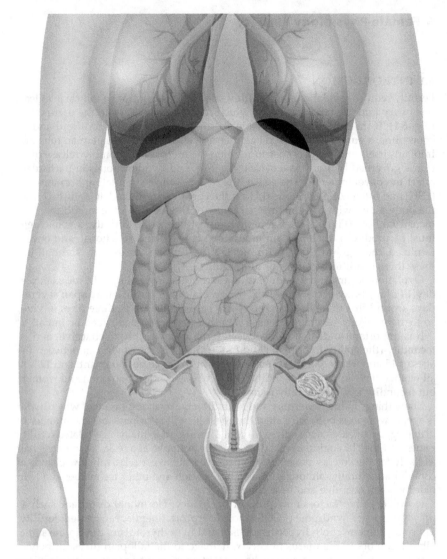

Figure 6.1 Ovaries and uterus.

among the brain, ovaries, and uterus. Thus, any drug that might interrupt or impede this communication could put fertility at risk. In addition, there are several ways these adverse effects may play out: drugs may impair ovulation, cause the endometrium to become unreceptive to a fertilized egg (Fig. 6.6), or in some way damage the uterus. Some of the drugs a woman could take that might affect fertility are described in the following.

6.2.1 Anticonvulsants

It is difficult to investigate whether and to what degree anticonvulsant drugs affect fertility, because epilepsy in and of itself appears to affect fertility and fecundity. Women with epilepsy report changes in seizure frequency and severity that track along with the hormonal changes at puberty, throughout each

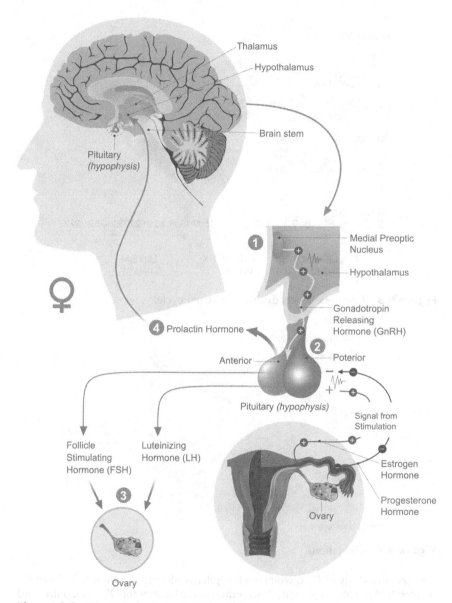

Figure 6.2 Pituitary hormones and ovaries.

menstrual cycle, and at menopause. Up to 40% of premenopausal women with epilepsy report catamenial epilepsy—that is, seizures that worsen with specific phases of the menstrual cycle, often but not exclusively around the time of ovulation.[2,3] This suggests that epilepsy affects the endocrine system in ways that may impede fertility. In a retrospective study of 1,000 women with epilepsy in the United States, 9.2% had tried to have a child but found they were infertile.[4] However, studies about infertility with epilepsy have been far from conclusive. One study of 197 women with epilepsy trying to become pregnant found that they were similarly likely to conceive as women without epilepsy.[5] In the

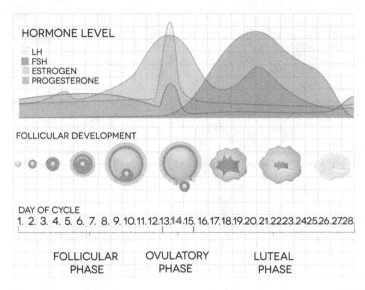

Figure 6.3 Hormone levels during menstrual cycle.

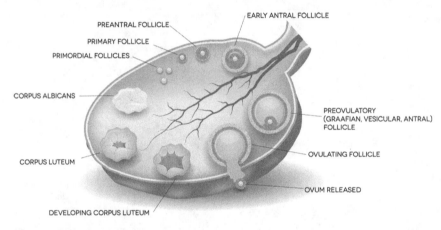

Figure 6.4 Ovulation.

retrospective study of 1,000 women with epilepsy already mentioned, 20.7% of the women had impaired fecundity.[4] According to the Centers for Disease Control and Prevention, in the time span from 2006 to 2010, the infertility rate among US women was 6.0%, and fecundity problems occurred in 12.0%.[6]

It is not clear whether anticonvulsant pharmacological therapy plays a role in the lower fertility and fecundity with epilepsy. Overall, the use of antiepileptic drugs has increased in recent years beyond the population of people with epilepsy. With newer drugs such as gabapentin and pregabalin available, more people with epilepsy may be prescribed drug therapy, and these anticonvulsants are prescribed for other indications as well, such as neuropathic pain treated by gabapentin or pregabalin. The prescribing patterns are shifting as well. Use of lamotrigine, levetiracetam, gabapentin, and pregabalin increased in use from 2001 to 2016, while valproate declined slightly.[7] In fact, in a registry study from Denmark, the

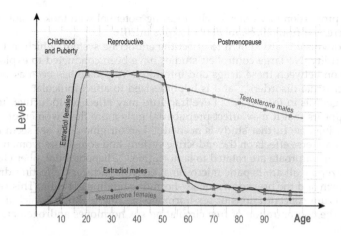

Figure 6.5 Age-related fertility.

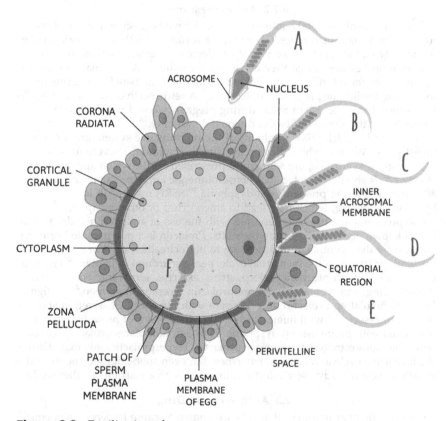

Figure 6.6 Fertilization of oocyte.

median proportion of women of childbearing potential who took an antiepileptic drug increased from 10.7% in 2001 to 27.1% in 2016.[7]

Anticonvulsants are known to affect the endocrine system, which in turn can affect fertility. No large controlled studies have been conducted to explore any association between these drugs and infertility. Valproate has been associated with menstrual disorders[8,9] and is hypothesized to affect follicular steroidogenesis in the ovary.[10] Levetiracetum may affect reproductive function, but it appears that it may affect prepubertal girls more than women of childbearing age; further study is needed.[10] Lamotrigine does not seem to have any clear adverse effects on the endocrine system, and sometimes women who do not tolerate valproate are rotated to lamotrigine.[11] Phenobarbital, phenytoin, and carbamazepine all are hepatic microsomal enzyme inducers, affecting drug metabolism and the breakdown of sex-hormone-binding globulin. This results in higher serum concentrations of sex-hormone-binding globulin, which in turn reduces the concentration of free circulating sex hormones androgen and estrogen.

In addition, some anticonvulsants may interact with oral contraceptives. Many antiepileptic drugs are human teratogens.[12] Despite these concerns, there have been few studies exploring the use of anticonvulsants during pregnancy.

6.2.2 Antidepressants

Selective serotonin reuptake inhibitors (SSRIs) have been shown to affect the function of the fallopian tubes.[13] However, a review of SSRIs and female fertility found only equivocal evidence that SSRIs decreased fertility, although a few studies found evidence that they decreased fecundity.[14] A systematic review of the literature for infertile women taking SSRIs found no benefits in terms of improving fertility or pregnancy outcomes.[15] A retrospective analysis of 920,620 births found that the use of SSRIs during pregnancy could not be statistically associated with rates of stillbirth or neonatal mortality.[16]

The ways in which SSRIs might affect the reproductive system are not well investigated. Women who suffer premenstrual dysphoria, a severe form of premenstrual syndrome, are sometimes prescribed SSRIs, which appears to be effective.[17] It should be noted that nonserotonergic antidepressants are not effective for treating premenstrual dysphoria.[18] SSRIs may result in hyperprolactinemia, which may lead to amenorrhea or other symptoms.[19] Hyperprolactinemia has been reported with the use of antipsychotic drugs, but less is known about its induction by SSRIs. Prolactin is a polypeptide hormone secreted by the hypothalamus according to both circadian rhythm and fluctuations based on the menstrual cycle, peaking around ovulation.[20] Prolactin levels increase with pregnancy, sleep, stress, exercise, sexual intercourse, and breastfeeding, and women have more prolactin receptors than men.[21,22] High levels of circulating prolactin may interfere with LH and FSH secretion and have been associated with infertility.[23,24] Abnormal levels of prolactin have been associated with premenstrual dysphoria. Estrogen modulates prolactin release, and exogenous estrogen can increase prolactin. Hyperprolactinemia may disrupt the menstrual cycle and cause amenorrhea, but even subtle derangements in the monthly cycle can adversely affect fertility.[19] This area warrants further study.

6.2.3 Antipsychotic Drugs

Antipsychotic pharmacological agents are known to cause hyperprolactinemia, which disrupts and may even stop the menstrual cycle, and in that way they impair fertility.[25] In a nested case-control study of 1,215 women with schizophrenia, those taking risperidone had four times the increase in prolactin levels as those who were not taking risperidone.[25] In a systematic review of antipsychotic agents and possible associations with female fertility ($n = 78$ studies), antipsychotic agents were found to affect the menstrual cycle in women and thus exert an adverse effect on fertility.[26] It is hypothesized that conventional antipsychotics inhibit dopamine at the D_2 receptors located in the hypothalamus (Fig. 6.7) and, in that way, increase prolactin secretion . Indeed, dopamine is likely the main inhibitory factor for prolactin. Many antipsychotic drugs nonselectively block D_2 receptors in all parts of the brain. Newer antipsychotic agents have less effect on prolactin levels.[26]

The study of antipsychotic drugs on fertility is confounded by the fact that women with schizophrenia have higher infertility rates than women without schizophrenia, apart from any pharmacological treatments. This does not negate the fact that antipsychotics can induce hyperprolactinemia, but rather suggests that antipsychotics may only be exacerbating preexisting underlying infertility. However, it is not clear why psychiatric disease would lead to hyperprolactinemia, although one hypothesis has put forth that the extreme stress of living with severe mental illness might lead to endocrine disruptions. Antipsychotic drugs clearly play a role in endocrine perturbations; typical antipsychotic treatment of three to nine weeks has been shown to raise prolactin levels 10 times over baseline values; tolerance develops over time and circulating prolactin levels decrease somewhat, but they remain at higher levels than normal.[26] Furthermore, the increase in circulating prolactin occurs in a dose-dependent fashion.[27] Atypical antipsychotics do not seem to elevate circulating levels of prolactin, with the exception of risperidone.[28] In fact, 72%–100% of women taking risperidone have hyperprolactinemia,[29,30] but only 1%–10% develop amenorrhea, while 48% have abnormal menstrual cycles.[31]

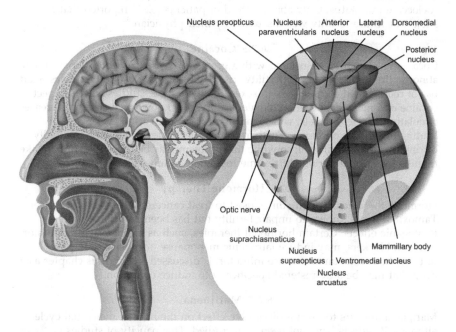

Figure 6.7 Nuclei of the hypothalamus.

6.2.4 Chemotherapeutic Agents

The oocytes, or cells of the ovaries, are rapidly dividing cells that are often destroyed by chemotherapeutic agents. In fact, some chemotherapeutic regimens will induce iatrogenic menopause by shutting down the ovaries. Many chemotherapeutic agents are known to affect female fertility, including, but not limited to, the family of platin drugs, doxorubicin, procarbazine, vinblastine, and vincristine. High doses, lengthier treatments, and the use of concomitant radiation can also increase the risk of infertility. Note that thalidomide, a drug known to cause severe birth defects, is sometimes used in chemotherapeutic regimens but should not be used in women who are or might become pregnant. (Sexually active patients of childbearing potential undergoing chemotherapy with either thalidomide or the related drug lenalidomide are asked to use two types of birth control throughout the duration of treatment.) Some patients can undergo chemotherapy and emerge with the ability to conceive and carry a child to term. In general, younger age at the time of chemotherapy can improve the odds of being fertile after treatment, as can certain types of treatment rather than others.

After a premenstrual girl undergoes chemotherapy, she may have menstrual periods at the appropriate age, but she is at risk for early menopause, defined as menopause that starts before age 40. It must be noted in this connection that one can be infertile and still have regular menstrual periods. The long-term effects of pediatric chemotherapy in cancer survivors are not well studied.

Patients considering chemotherapy who wish to have a child (or more children) should discuss these concerns with their physician before treatment. Healthcare professionals do not always initiate these discussions, and oncologists may have such intense focus on cancer treatment that they fail to consider fertility preservation in their patients. In a cross-sectional study at a tertiary care center in 2019, only 32% of patients understood that cancer treatments might affect their fertility, and this awareness was significantly lower in patients of low socioeconomic status. Only about a third of patients (32%) reported that the subject of future fertility was discussed by their physician.[32]

6.2.5 Cocaine

Cocaine use has been associated with a greatly increased risk of fallopian tubal abnormality that can cause infertility (risk ratio, 11.1).[33] Despite the widespread use of recreational cocaine in both sexes, there are more studies on the subject of cocaine and male fertility than female fertility. It is known that cocaine increases circulating prolactin levels, which can disrupt the menstrual cycle, possibly leading to fertility problems. Furthermore, prolactin levels increase markedly during cocaine withdrawal, so while this problem may be reversible, it may take months for normal periods to resume following cocaine detoxification.

6.2.6 Hormone Therapy

Hormone therapies are sometimes used to treat cancers, such as breast cancer. Tamoxifen is not known to impair fertility but has been associated with teratogenic effects. Certain hormonal therapies, such as aromatase inhibitors or estrogen blockers, may induce iatrogenic menopause or otherwise impair fertility. Letrozole, an aromatase inhibitor, is discussed later in this chapter as a drug that may be administered specifically to induce ovulation.

6.2.7 Marijuana

Marijuana appears to exert a disruptive effect on the normal menstrual cycle, although its effects have not been well studied. This paucity of studies is surprising, as marijuana use is common and becoming more mainstream. Regular marijuana use in women is associated with a higher risk of anovulatory cycles, which are associated with infertility.[33,34] Occasional marijuana use can delay normal ovulation.[34] Delta-9 tetrahydrocannabinol, the psychoactive constituent of marijuana, inhibits the release of gonadotropin-releasing hormone and thyrotropin-releasing hormone in the brain's hypothalamus, which leads to less production of prolactin, FSH, and LH from the pituitary gland, all of which regulate egg production.[35]

6.2.8 Nonsteroidal Anti-Inflammatory Drugs

Nonsteroidal anti-inflammatory drugs (NSAIDs) are a broad class of drugs that includes some of the most frequently consumed over-the-counter and prescription analgesics in the world. While not thoroughly elucidated, it appears that a course of NSAIDs over at least 10 days may reduce progesterone levels and impede ovulation. In one study, 49 women of childbearing age were randomized to receive 100 mg/day diclofenac, 500 mg of naproxen twice a day, 90 mg/day etoricoxib, or placebo. Women initiated treatment on day 10 of the menstrual cycle and took the drugs every day for 10 days. All participants in the placebo group ovulated during their next cycle, while only 25% of the diclofenac group, 75% of the naproxen group, and 66% of the etoricoxib group experienced a normal ovulation. All women taking an NSAID in this study had lower progesterone levels than in the placebo group. It should be noted that all participants returned to normal ovulatory cycles one month after discontinuing NSAID therapy.[36]

6.2.9 Opioids

Both prescription and illicit opioids affect the endocrine system and disrupt activity in the hypothalamus and pituitary gland, leading to decreases in FSH and LH, which can disrupt the normal menstrual cycle. While men consume more illicit opioids than women, women are more likely than men to have a prescription for opioids.[37] In a matched-cohort study of 44,260 women of childbearing age (18–55 years) prescribed an opioid for musculoskeletal pain from 2002 to 2013, long-term use of an opioid analgesic was associated with alterations in menstrual cycle (hazard ratio, 1.13) and an increased risk of menopause (hazard ratio, 1.16) but not infertility (hazard ratio, 0.82).[38] A systematic literature review reports that 23%–71% of women taking oral or intrathecal opioids long-term experienced amenorrhea.[39]

Opioid-induced androgen deficiency is a condition associated with decreased fertility, along with other symptoms such as fatigue, decreased muscle mass, weight gain, and osteoporosis.[40] In a study of 47 women taking chronic opioid therapy (age range, 30–75 years) with a sustained-action opioid analgesic compared to 68 women who did not take opioids, it was found that testosterone, estradiol, and dehydroepiandrosterone sulfate were 48%–57% lower in women with intact ovaries who took opioids, and premenopausal women taking opioids had 30% lower levels of LH and FSH.[41] This suggests that their fertility would be reduced. While opioid-associated endocrinopathies are prevalent in both men and women, they are often not reported by patients nor observed by clinicians, and as a result, they are likely underestimated and in turn not well investigated.[42]

6.2.10 Steroids

Those who participate in competitive sports or regular high-intensity exercise exhibit a rate of hypothalamic amenorrhea of up to 40%, much higher than occurs in those who do not exercise at this intensity.[43] Anovulation rates are also higher in the more athletic than the more sedentary.[43] For about half a century, this finding was explained by the "critical fat" hypothesis, which maintained that a certain amount of body fat is needed in order to begin and maintain reproductive function. This has given way in recent years to the "metabolic fuel" theory, which states that bodies require a specific amount of energy, rather than adiposity, to maintain reproductive function.[44] Moderate exercise seems to enhance fertility, while very intense prolonged exercise impairs it.

A small number of athletes take performance-enhancing drugs, the most common of which are anabolic-androgenic steroids (often called just steroids). It has been estimated that about 2% of female athletes take steroids,[45] which can lead to dysmenorrhea, amenorrhea, and anovulation. In prepubescent athletes, the use of these agents can delay menarche.[46] Steroids can lead to hypogonadism.They can lead to male pattern baldness, hirsutism, and clitoral hypertrophy, which may be irreversible. However, studies of the effects of steroids on fertility are confounded by the infertility induced by extreme exercise programs apart from any drugs.

6.2.11 Thyroid Medications

Thyroid function (Fig. 6.8) is closely related to the endocrine system, and in that way plays a crucial role in reproductive health. Thyroid disorders are five to eight times more prevalent in women than men and can affect fertility.

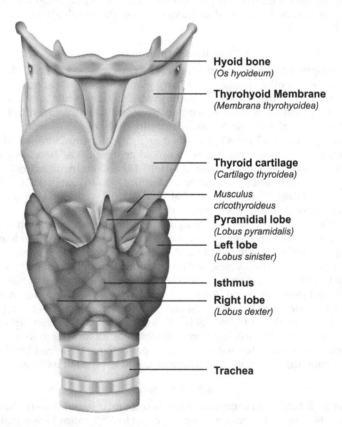

Figure 6.8 Location of thyroid gland.

Both hyperthyroid and hypothyroid disorders can cause abnormal menstrual cycles, but hypothyroidism is more frequently associated with female infertility.

Hyperthyroidism (Fig. 6.9), which occurs in about 5% of women of childbearing age, is typically treated with thyroxine, a hormonal treatment that may normalize menstrual cycles and improve fertility. Untreated, an overactive thyroid may result in anxiety, insomnia, weight loss or inability to gain weight, and fewer or lighter menstrual periods.

Hypothyroidism, which occurs in about 2%–4% of women of childbearing age, can cause infertility in and of itself.[47] A slow thyroid is typically treated with thyroid-stimulating hormone and prolactin. Untreated hypothyroidism may result in weight gain, constipation, and more frequent and heavier periods, which

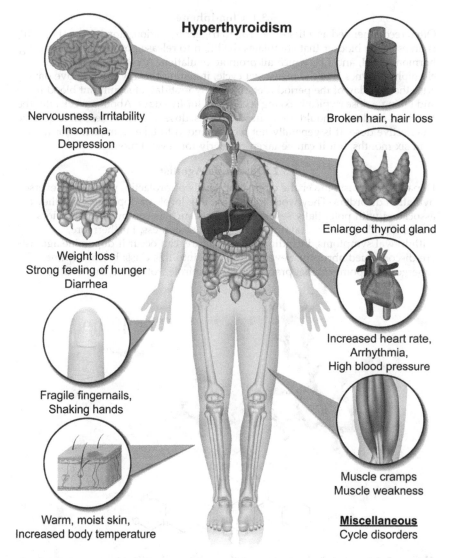

Hyperthyroidism

Nervousness, Irritability Insomnia, Depression

Broken hair, hair loss

Enlarged thyroid gland

Weight loss Strong feeling of hunger Diarrhea

Increased heart rate, Arrhythmia, High blood pressure

Fragile fingernails, Shaking hands

Muscle cramps Muscle weakness

Warm, moist skin, Increased body temperature

Miscellaneous Cycle disorders

Figure 6.9 Hyperthyroidism.

are often associated with infertility. In a study of 394 infertile women, 24% were found to have hypothyroidism. When this was treated pharmacologically, 77% conceived within one year.[47] Hypothyroidism is associated with hyperprolactinemia; it can be effectively treated pharmacologically, which can often restore fertility.

6.3 DRUGS THAT PROMOTE FERTILITY

Obviously, ovulation is crucial for fertility. Several drugs can help stimulate regular ovulation and, in that way, improve the chances of becoming pregnant. These are drugs that would be taken specifically to encourage ovulation, as opposed to the drugs already described, which would be taken for some other purpose but might have side effects that alter fertility and fecundity.

6.3.1 Clomiphene

Often recommended as a first-line drug to trigger ovulation, clomiphene (Fig. 6.10) is an estrogen blocker that stimulates the brain to release gonadotropin-releasing hormone, FSH, and LH, which all promote ovulation. A patient prescribed clomiphene syncs it to the menstrual cycle, initiating therapy three to five days after the first day of the period (determined by first day of significant blood flow) and taking a dose (typically 50 mg) every day for five days. About a week after the last dose, ovulation should occur. In some cases, dose titration is needed to achieve an effective dose. It is generally not recommended to take clomiphene for more than six months, but it can be taken regularly for several months.

6.3.2 Dopamine Agonists

Dopamine agonists lower the circulating levels of prolactin, which can reverse ovulation disorders. They work to increase low levels of dopamine, but they are associated with potentially severe side effects such as compulsive behaviors, mental health disorders, fainting, and abrupt sleepiness. Furthermore, withdrawal symptoms, including kidney failure, can occur if dopamine agonists are discontinued abruptly. Several drugs fall into this class: bromocriptine, cabergoline, apomorphine, pramipexole, ropinirole, rotigotine, and others.

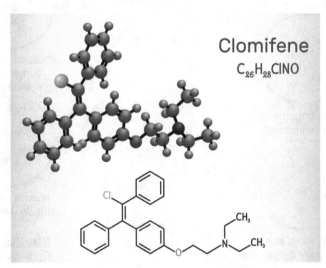

Figure 6.10 Chemical structure of clomiphene (clomifene).

6.3.3 Metformin (Glucophage)

PCOS is a condition that results in anovulation and is frequently comorbid with obesity. Metformin is a well-known antidiabetes drug that helps manage glycemic control and combats insulin resistance. It is thought that high levels of insulin the body may cause PCOS and secondary anovulation. Metformin can help better manage serum insulin and, in that way, can promote ovulation. However, it is not generally advocated that metformin be taken solely for the purpose of stimulating ovulation, although it may have that effect in patients taking it for insulin resistance.

However, metformin is sometimes taken with clomiphene to treat infertility due to anovulation. In a case series of 48 women with PCOS taking metformin initiated at 500 mg twice daily for six weeks and then increased to up to 500 mg as needed with adjunctive 50 mg clomiphene, likewise as needed, until ovulation occurred, 40% of participants had spontaneous menses and evidence of ovulation with metformin alone, and another 31% experienced the same with metformin plus clomiphene. In that study, 42% of women conceived (median time to conception, 3 months) but 35% had spontaneous abortions.[48] In a study of infertile women in Asia, 68% of participants treated with both metformin and clomiphene ovulated, 21% became pregnant, and 18.4% had a live birth.[49]

6.3.4 Gonadotropins

Gonadotropins are hormones that work to stimulate the ovaries and bring about ovulation. These drugs may be administered as an injection or intranasally and are typically used as a second- or third-line treatment. They are marketed under trade names including Novarel, Ovidrel, Pregnyl, Profasi, Menopur, Metrodin, Pergonal, Repronex, Factrel, Lutrepulse, Lupron, Synarel, Zoladex, Antagon, and Cetrotide. They are sometimes used in combination with each other or with other types of agents.

6.3.5 Letrozole

Letrozole is an aromatase inhibitor that can trigger ovulation in premenopausal women. Estrogen in the body is produced in part by the action of the aromatase enzyme system; letrozole inhibits this enzyme and, in that way, lowers estrogen levels. This triggers the pituitary gland to produce more FSH and LH, which stimulate ovulation. Letrozole is sometimes prescribed to treat anovulation in women with PCOS or morbid obesity.

It may also be prescribed to those who are undergoing fertility treatments. In recipients who ovulate naturally, letrozole can be used for superovulation, or the production of multiple eggs at once, which can be important for artificial insemination. Those who take letrozole have a 10% chance of conceiving twin.

Letrozole is also prescribed for treating postmenopausal women with hormone-receptor-positive, early-stage breast cancer to reduce the risk of recurrence. It may also be prescribed to postmenopausal women with advanced breast cancer or metastatic hormone-receptor-positive forms of breast cancer. (Letrozole is not effective against hormone-receptor-negative breast cancer.)

6.4 CONCLUSION

The monthly release of an egg is governed by hormonal cycles and results in either fertilization (when egg encounters sperm) or menstruation, where the unfertilized egg is flushed out of the body. Fertility is governed by the monthly cycle, with ovulation and optimal fertility occurring around day 14 of a 28-day cycle. Age can affect fertility in the form of the "biological clock", and even those who still produce eggs are less fertile as they age. Other factors, such as hormonal conditions, diseases, and fibroid tumors in the uterus or cervix are among the things that can adversely affect fertility. Certain drugs may also adversely affect fertility, such as antiepileptics, antidepressants, and antipsychotics, although there are not many studies on the subject. Chemotherapy may affect fertility, and those undergoing such treatments may wish to discuss options for preserving eggs. Likewise, certain drugs, such as clomiphene, may enhance fertility.

REFERENCES

1. Centers for Disease Control and Prevention. Infertility FAQs. *Reproductive Health*. https://www.cdc.gov/reproductivehealth/infertility/index.htm. Published 2020. Accessed August 27, 2020.

2. Morrell MJ. Epilepsy in women: The science of why it is special. *Neurology*. 1999;53(4 Suppl 1):S42–S48.

3. Maguire MJ, Nevitt SJ. Treatments for seizures in catamenial (menstrual-related) epilepsy. *Cochrane Database Syst Rev*. 2019;10(10):Cd013225.

4. MacEachern DB, Mandle HB, Herzog AG. Infertility, impaired fecundity, and live birth/pregnancy ratio in women with epilepsy in the USA: Findings of the epilepsy birth control registry. *Epilepsia*. 2019;60(9):1993–1998.

5. Pennell PB, French JA, Harden CL, et al. Fertility and birth outcomes in women with epilepsy seeking pregnancy. *JAMA Neurol*. 2018;75(8):962–969.

6. Chandra A, Copen CE, Stephen EH. Infertility and impaired fecundity in the united states, 1982–2010: Data from the national survey of family growth. *Natl Health Stat Report*. 2013(67):1–18, 11 p following 19.

7. Daugaard CA, Sun Y, Dreier JW, Christensen J. Use of antiepileptic drugs in women of fertile age. *Dan Med J*. 2019;66(8).

8. Isojärvi JI, Laatikainen TJ, Pakarinen AJ, Juntunen KT, Myllylä VV. Polycystic ovaries and hyperandrogenism in women taking valproate for epilepsy. *N Engl J Med*. 1993;329(19):1383–1388.

9. Isojärvi JI, Taubøll E, Herzog AG. Effect of antiepileptic drugs on reproductive endocrine function in individuals with epilepsy. *CNS drugs*. 2005;19(3):207–223.

10. Svalheim S, Sveberg L, Mochol M, Taubøll E. Interactions between antiepileptic drugs and hormones. *Seizure*. 2015;28:12–17.

11. Isojärvi JI, Rättyä J, Myllylä VV, et al. Valproate, lamotrigine, and insulin-mediated risks in women with epilepsy. *Ann Neurol*. 1998;43(4):446–451.

12. Morrell MJ. The new antiepileptic drugs and women: Efficacy, reproductive health, pregnancy, and fetal outcome. *Epilepsia*. 1996;37 Suppl 6:S34–S44.

13. Milosavljević MN, Janković SV, Janković SM, et al. Effects of selective serotonin reuptake inhibitors on motility of isolated fallopian tube. *Clin Exp Pharmacol Physiol.* 2019;46(8):780–787.

14. Sylvester C, Menke M, Gopalan P. Selective serotonin reuptake inhibitors and fertility: Considerations for couples trying to conceive. *Harv Rev Psychiatry.* 2019;27(2):108–118.

15. Domar AD, Moragianni VA, Ryley DA, Urato AC. The risks of selective serotonin reuptake inhibitor use in infertile women: A review of the impact on fertility, pregnancy, neonatal health and beyond. *Hum Reprod.* 2013;28(1):160–171.

16. Jimenez-Solem E, Andersen JT, Petersen M, et al. SSRI use during pregnancy and risk of stillbirth and neonatal mortality. *Am J Psychiatry.* 2013;170(3):299–304.

17. Eriksson E. Serotonin reuptake inhibitors for the treatment of premenstrual dysphoria. *Int Clin Psychopharmacol.* 1999;14(Suppl 2):S27–S33.

18. Eriksson E, Andersch B, Ho HP, Landén M, Sundblad C. Diagnosis and treatment of premenstrual dysphoria. *J Clin Psychiatr.* 2002;63(Suppl 7): 16–23.

19. Emiliano AB, Fudge JL. From galactorrhea to osteopenia: Rethinking serotonin-prolactin interactions. *Neuropsychopharmacology.* 2004;29(5):833–846.

20. Seppälä M. Prolactin and female reproduction. *Ann Clin Res.* 1978; 10(3):164–170.

21. Yazigi RA, Quintero CH, Salameh WA. Prolactin disorders. *Fertil Steril.* 1997;67(2):215–225.

22. Chiu S, Wise PM. Prolactin receptor mRNA localization in the hypothalamus by in situ hybridization. *J Neuroendocrinol.* 1994;6(2): 191–199.

23. Gómez F, Reyes FI, Faiman C. Nonpuerperal galactorrhea and hyperprolactinemia. Clinical findings, endocrine features and therapeutic responses in 56 cases. *Am J Med.* 1977;62(5):648–660.

24. Katz E, Adashi EY. Hyperprolactinemic disorders. *Clin Obstet Gynecol.* 1990;33(3):622–639.

25. Chen H, Qian M, Shen X, et al. Risk factors for medication-induced amenorrhea in first-episode female Chinese patients with schizophrenia treated with risperidone. *Shanghai Arch Psychiatry.* 2013;25(1):40–47.

26. Bargiota SI, Bonotis KS, Messinis IE, Angelopoulos NV. The Effects of Antipsychotics on Prolactin Levels and Women's Menstruation. *Schizophr Res Treatment.* 2013;2013:502697.

27. Spitzer M, Sajjad R, Benjamin F. Pattern of development of hyperprolactinemia after initiation of haloperidol therapy. *Obstet Gynecol.* 1998;91(5):693–695.

28. Knegtering H, van der Molen A, Castelein S, Kluiter H, van den Bosch R. What are the effects of antipsychotics on sexual dysfunctions and endocrine funcioning? *Psychoneuroendocrinology.* 2013;28:109–203.

29. Dickson R, Glazer W. Neuroleptic-induced hyperprolactenemia. *Schizophr Res.* 1999;35:575–586.

30. Meaney A, O'Keane V. Prolactin and schizophrenia: Clinical consequences of hyperprolactinaemia. *Life Sci.* 2002;71(9):979–992.

31. Kinon B, Gilmore J, Liu H, Halbreich U. Prevalence of hyperprolactinemia in schizophrenic patients treated with conventional antipsychotic medications or risperidone. *Psychoneuroendocrinology.* 2003;28:55–68.

32. Mahey R, Kandpal S, Gupta M, Vanamail P, Bhatla N, Malhotra N. Knowledge and awareness about fertility preservation among female patients with cancer: A cross-sectional study. *Obstet Gynecol Sci.* 2020;63(4):480–489.

33. Mueller BA, Daling JR, Weiss NS, Moore DE. Recreational drug use and the risk of primary infertility. *Epidemiology.* 1990;1(3):195–200.

34. Jukic AM, Weinberg CR, Baird DD, Wilcox AJ. Lifestyle and reproductive factors associated with follicular phase length. *J Womens Health (Larchmt).* 2007;16(9):1340–1347.

35. Brents LK. Marijuana, the endocannabinoid system and the female reproductive system. *Yale J Biol Med.* 2016;89(2):175–191.

36. Harrison P. NSAIDs dramatically reduce ovulation with consistent use. Medscape. *Medscape Medical News.* https://www.medscape.com/viewarticle/846552?nlid=83026_2042. Published 2015. Accessed August 28, 2020.

37. How the U.S. opioid crisis affects women and infertility medicine. IRMS. https://www.sbivf.com/blog/how-the-u-s-opioid-crisis-affects-women-and-infertility-medicine/. Published 2017. Accessed August29, 2020.

38. Richardson E, Bedson J, Chen Y, Lacey R, Dunn KM. Increased risk of reproductive dysfunction in women prescribed long-term opioids for musculoskeletal pain: A matched cohort study in the Clinical Practice Research Datalink. *Eur J Pain.* 2018;22(9):1701–1708.

39. Wersocki E, Bedson J, Chen Y, LeResche L, Dunn KM. Comprehensive systematic review of long-term opioids in women with chronic noncancer pain and associated reproductive dysfunction (hypothalamic-pituitary-gonadal axis disruption). *Pain.* 2017;158(1):8–16.

40. Smith HS, Elliott JA. Opioid-induced androgen deficiency (OPIAD). *Pain Physician*. 2012;15(3 Suppl):Es145–Es156.

41. Daniell HW. Opioid endocrinopathy in women consuming prescribed sustained-action opioids for control of nonmalignant pain. *J Pain*. 2008;9(1):28–36.

42. Brennan MJ. The effect of opioid therapy on endocrine function. *Am J Med*. 2013;126(3 Suppl 1):S12–S18.

43. Scheid JL, De Souza MJ. Menstrual irregularities and energy deficiency in physically active women: The role of ghrelin, PYY and adipocytokines. *Med Sport Sci*. 2010;55:82–102.

44. Mircea CN, Lujan ME, Pierson RA. Metabolic fuel and clinical implications for female reproduction. *J Obstet Gynaecol Can*. 2007;29(11):887–902.

45. Sagoe D, Molde H, Andreassen CS, Torsheim T, Pallesen S. The global epidemiology of anabolic-androgenic steroid use: A meta-analysis and meta-regression analysis. *Ann Epidemiol*. 2014;24(5):383–398.

46. La Vignera S, Condorelli RA, Cannarella R, Duca Y, Calogero AE. Sport, doping and female fertility. *Reprod Biol Endocrinol*. 2018;16(1):108–108.

47. Verma I, Sood R, Juneja S, Kaur S. Prevalence of hypothyroidism in infertile women and evaluation of response of treatment for hypothyroidism on infertility. *Int J Appl Basic Med Res*. 2012;2(1):17–19.

48. Heard MJ, Pierce A, Carson SA, Buster JE. Pregnancies following use of metformin for ovulation induction in patients with polycystic ovary syndrome. *Fertil Steril*. 2002;77(4):669–673.

49. Zain MM, Jamaluddin R, Ibrahim A, Norman RJ. Comparison of clomiphene citrate, metformin, or the combination of both for first-line ovulation induction, achievement of pregnancy, and live birth in Asian women with polycystic ovary syndrome: A randomized controlled trial. *Fertil Steril*. 2009;91(2):514–521.

7 Oral Contraception and Hormone Replacement Therapy

7.1 INTRODUCTION

Barring injury or illness, males are always fertile, and while their fertility diminishes gradually with advancing age, even very old men may still father children. The oldest father on record fathered a child at age 96 (he passed away in 2020 at the age of 104). Females, on the other hand, have specific temporal limitations on their fertility. They are fertile only a few days each month, and this fertility lasts only a few decades of their lifetime. These seasons of fertility and infertility are regulated by complex hormonal interactions.

7.2 HORMONAL FACTORS

7.2.1 Oral Hormone Contraception

The menstrual cycle is divided into specific but somewhat overlapping phases (Fig. 7.1). The menstrual phase ('menstruation'), during which significant menstrual bleeding first occurs, is characterized by higher levels of the endogenous hormone progesterone. Progesterone inhibits the production of two important hormones: follicle-stimulating hormone (FSH) and luteinizing hormone (LH; Fig. 7.2). This would be day 1 of a normal 28-day cycle. Approaching the middle of the cycle, around day 10–12, progesterone levels decrease, allowing for an increase in FSH and LH. FSH and LH are necessary to stimulate the follicles (Fig. 7.3) to produce an egg and thicken the uterine walls in preparation for implantation of a fertilized egg. Just before mid-cycle, day 14, estrogen levels increase markedly (up to 800%) and, along with high levels of FSH and LH, trigger ovulation;[1] progesterone levels remain low. Once ovulation occurs, estrogen levels remain high and progesterone levels ramp up in preparation for an embryo. If a fertilized egg is implanted in the uterus, both progesterone and estrogen levels remain high to support the pregnancy and prevent another ovulation from occurring. If no egg is implanted in the uterus in the next two weeks, both progesterone and estrogen levels drop off sharply, the thickened surface of the uterine walls is sloughed off, and the next menstrual bleeding occurs. While the typical menstrual cycle in humans is 28 days, longer and shorter cycles are not unusual.

First developed in the 1950s and legally available in the United States in 1972, oral contraceptives are based on pharmacologically disrupting the hormonal system so that ovulation does not occur. Oral contraceptives rely on exogenous progesterone alone or in combination with estradiol to keep hormonal levels such that ovulation is prevented. There are essentially three types of oral contraceptives: the older estradiol-and-progesterone pill, a progesterone-only formulation, and a new extended-use pill. Some of these contraceptives offer three weeks of pills and a week of placebo; during the placebo period, bleeding occurs. It should be noted that although there is monthly bleeding, this is not a normal menstrual period, but rather breakthrough bleeding. In fact, contraceptives are sometimes prescribed for women with painful periods because they suppress normal menstruation. It is sometimes claimed that oral contraceptives "trick" the body into thinking it is already pregnant and thus preventing ovulation. This is not entirely accurate; the hormonal shifts induced by the pill mimic the prolonged "waiting period" following ovulation before a fertilized egg arrives in the womb.

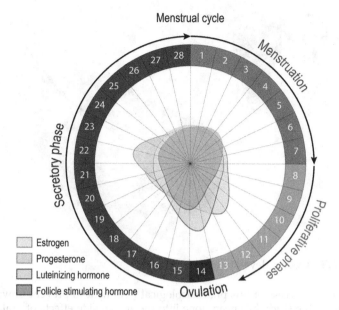

Figure 7.1 The menstrual cycle: hormones.

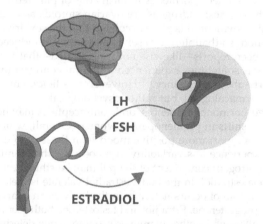

Figure 7.2 Luteinizing hormone (LH), follicle-stimulating hormone (FSH), and estradiol.

New formulations of oral contraceptives have been developed that reduce the number of periods experienced from one a month to four a year or even none. Indeed, oral contraceptives are sometimes used to manipulate the monthly cycle, for example to avoid having a period on vacation. From a medical point of view, such suppression is likely harmless, in that it only mimics the body's normal activities in anticipation of pregnancy.

Oral contraceptives are considered to be safe and effective for most women. With so-called perfect use, they are 99% effective in preventing pregnancy in one year. However, human error—usually forgetting to take the pill—accounts for the fact that real-world pregnancy rates can be as high as 9% for pill users in a year.[2]

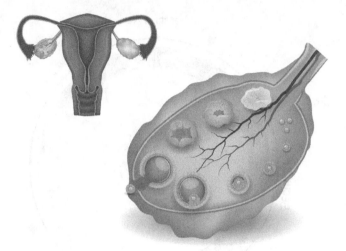

Figure 7.3   Oocyte development.

Despite the success of this pharmacological treatment to prevent unwanted pregnancy, little study has been done into the sexual side effects of oral contraceptives. Decreased libido has been reported in about 5%–10% of users. It is theorized that the synthetic hormones in birth control pills reduce androgen production by the ovaries; androgens are ovarian steroids associated with sex drive and sexual enjoyment. It is also speculated that the sex drive is hormonally synced to ovulation, so thatinterest in sex is lost with pill-induced anovulation. There have also been reports—likewise not well studied—that increased interest in sexual activity is sometimes experienced with oral contraceptives because there is less worry about pregnancy.[3,4] However, this libido-enhancing benefit would not be pharmacological, but rather psychological.

Because of their hormonal influences, oral contraceptives may help smooth out cycle-based mood shifts in premenopausal women. Overall, women who take oral contraceptives report more positive mood than those who do not take the pill,[5] and they experience less variability in mood over the month.[6]

The estradiol (estrogen) component of the pill may be ethinyl estradiol, estradiol valerate, or 17-beta-estradiol. In the body, estradiol valerate is metabolized to 17-beta-estradiol. Estradiol occurs naturally in the body as an ovarian steroid, like testosterone and progesterone. Estradiol and testosterone both seem to be crucial in sexual response, although in other mammals estradiol alone determines sexual response in females.[1] Ovarian-produced androgens fluctuate with the menstrual cycle, reaching the high point around mid-cycle with ovulation, but levels at the maximum point are only about 150% above low levels.[1]

The use of oral contraceptives results in lower levels of circulating androgens, estradiol, and progesterone and may inhibit the production of oxytocin.[4] Oxytocin (Fig. 7.4) is a hypothalamic peptide that is associated with pair-bonding behaviors and emotional connections. In humans, peripheral oxytocin levels are very high in new lovers and are thought to explain the strong attraction and affinity experienced in the "honeymoon phase" of romantic relationships. In testing, oxytocin reduces jealousy scores with respect to potential infidelity and increases interest in nulliparous women for infants and children. Oxytocin can also increase orgasmic intensity and the feelings of postcoital contentment and satisfaction. Thus, a decrease in oxytocin (as might occur with oral

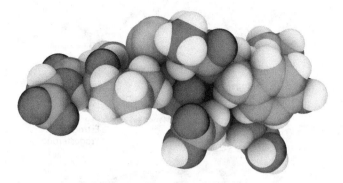

Figure 7.4   3-D representation of oxytocin molecule.

contraceptives) may increase jealousy responses, diminish orgasmic intensity, and decrease pair-bonding emotions in women.[4] In fact, the decrease in oxytocin brought about by hormonal contraceptives is thought to reduce pair-bonding in women.[4]

Other side effects reported for oral contraceptives include vaginal dryness, dyspareunia, urinary incontinence, vestibulodynia, and interstitial cystitis. Physical changes to the genitals, such as thickening of the labia minora, have occurred with oral contraceptives and may result in diminished sexual pleasure.[4] In contrast to oral contraception, the use of contraceptive patches and vaginal rings appears to have less effect on sexuality.[4] In a study comparing the use of hormonal and nonhormonal contraception in females ($n = 1,101$), hormonal contraceptives were associated with less frequent sexual activity, less sexual arousal, increased vaginal dryness, less pleasure in sex, and less frequent or less intense orgasm.[7] It should be noted here that frequency of sexual activity may not be a reliable marker for sexual desire, because women are able to have sex at any point during their menstrual cycle.

The European Society of Sexual Medicine has declared that there is insufficient evidence to state that hormonal contraception can increase sexual dysfunction, but it urges further study, as a minority of women taking these drugs report sexual problems.[8] A narrative review of the literature reports that sexual dysfunction occurs in a subset of women taking oral contraceptives.[9] In a survey conducted in Germany ($n = 1,219$) about female sexual dysfunction, it was found that about one-third of women (32.4%) had some degree of sexual dysfunction: problems were centered around orgasm (8.7%), sexual desire (5.8%), sexual satisfaction (2.6%), lubrication (1.2%), pain during intercourse (1.1%), and lack of sexual arousal (1.0%). The use of a hormonal contraceptive was associated with lower desire and lower arousal scores than in women who did not take hormonal contraceptives.[10]

Drug-drug interactions are possible with oral contraceptives, and if such interactions decrease the effectiveness of progesterone, they can result in ovulation and possible pregnancy even in those who take the pill as prescribed. Among the drugs that may interact with the progesterone component of oral contraceptives are antiretroviral agents.

7.2.2 Hormone Replacement Therapy

The same hormonal fluctuations that govern ovulation and fertility change as the supply of viable eggs runs out and ovarian hormone production diminishes

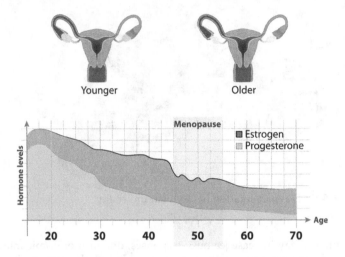

Figure 7.5 Estrogen and progesterone levels with age.

(Fig. 7.5). Menopause typically occurs at around age 50 naturally, or may occur earlier as iatrogenic menopause, induced by chemotherapy, oophorectomy, or another intervention. In perimenopause, endogenous estrogen levels decrease, which may cause physical, mental, and sexual symptoms (Fig. 7.6). Indeed, menopause is experienced by many as a major physical and emotional life transition that may be marked by introspection, crisis, distress, or renewed engagement in life. The most visible symptom of menopause is the cessation of menstruation, which marks the end of the normal ability to conceive a child. Normal menopause can cause a number of mild to severe symptoms that may be perceived as annoying to debilitating: vasomotor dysfunction, urinary symptoms, atrophic vaginitis, and sexual problems.[11] Postmenopausal women report high levels of sexual dysfunction, and these rates vary by country, ranging from about 30% in the United States[12] to 68% in Lithuania[13] and 85.2% in Malaysia.[14] Furthermore, sexual dysfunction is associated with quality of life.[15] Menopausal symptoms can last for years.

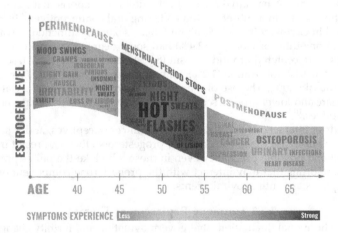

Figure 7.6 Symptoms associated with menopause.

Hormone replacement therapy (HRT) encompasses a variety of pharmacological approaches to treat some of the symptoms of menopause. However, HRT has been associated with serious risks such as increased rates of blood clots, myocardial infarction, stroke, breast cancer, and gallbladder disease. It may involve estrogen only or a combination of estrogen and progestin (a synthetic version of progesterone).[16] Estrogen-only therapy is associated with a risk for endometrial cancer, so combination types of HRT are prescribed for women who have an intact uterus. In women who have had a hysterectomy, estrogen-only HRT may be prescribed. In women over age 65 taking HRT, there is an elevated risk for dementia, but there are also certain benefits associated with it, including decreased incidence of osteoporosis and lower risk for vertebral fractures. Estrogen-only HRT may protect against breast cancer, but combination HRT increases the risk.[16]

The use of HRT is associated with a lower risk for sexual dysfunction.[17] Tibolone (Fig. 7.7) likewise reduces sexual dysfunction and may be an alternative to HRT in women who have sexual symptoms with HRT.[18] Tibolone is a selective tissue estrogenic activity regulator that stimulates hormonal receptors in specific tissues. Since estrogen's effects occur mainly in the brain, bones, and vagina, tibolone's selective action may be advantageous. Tibolone has been reported to improve mood, sharpen cognition, decrease neuroinflammation, reduce vasomotor symptoms, and improve sexual response in postmenopausal women.[19] It may also lower high-density lipoprotein levels.[20]

Menopause is somewhat of a human anomaly, as it does not occur in most animal species, where females remain fertile throughout the course of their lifetime. Among mammals, pilot whales and a few types of Asian elephants are the only other species with menopause.[21] Other primates, for example, appear to be fertile to the end of life, although there is some speculation that certain female primates do experience some form of natural diminishment of fertility that may parallel menopause in humans.[22] For example, gradually increasing cycle length has been observed in aging chimpanzees, and female chimpanzees may cease all cycles as a result of illness.[22]

The unusual nature of human menopause has led to speculation as to why reproductive age is so limited for women. The adaptive rationale behind human menopause seems puzzling, in that from an evolutionary standpoint it would seem beneficial for long-lived females to be able to pass on good genes associated with longevity, and so they ought to be fertile for as long as possible. From an evolutionary standpoint, it seems counterintuitive that females live decades past their reproductive potential. In populations of natural fertility, females typically have their first child at age 19 and their last child at age 38. Fertility decreases sharply at around age 40, but menopause does not occur for another 10 years.

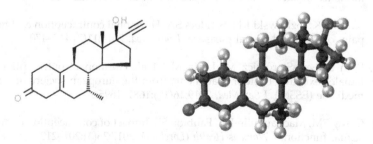

Figure 7.7 Chemical structure of tibolone.

Thus, females find it increasingly more difficult to have a child after the age of 40, but natural pregnancy is often still possible. With medical interventions such as *in vitro* fertilization, pregnancy is possible beyond menopause. However, natural pregnancies are difficult after 40 and impossible after menopause.

The evolutionary reasons that women might live decades beyond their natural fertility have never been entirely elucidated. The "grandmother hypothesis" has been advanced to explain human menopause, in which in social units (extended families, communities), women stop reproducing at around the same time as the next generation of women is ready to reproduce.[23] But this is by no means a definitive explanation.

7.3 CONCLUSION

Oral hormone contraception can help prevent pregnancy by altering hormonal levels to prevent the egg from being implanted in the uterus. While this form of contraception is effective, diminished libido is reported in about 5% to 10% of users. Menopause occurs when ovulation ceases and this may occur naturally with age or iatrogenically, such as after an oophorectomy or as a result of chemotherapy. Menopause is associated with mild to severe physical, mental, and sexual symptoms which might be treated with hormone replacement therapy. Sexual dysfunction is frequently reported as women enter menopause.

REFERENCES

1. Cappelletti M, Wallen K. Increasing women's sexual desire: The comparative effectiveness of estrogens and androgens. *Horm Behav*. 2016;78:178–193.

2. Cooper D, Mahdy H. *Oral conctraceptive pills*. Treasure Island, FL: StatPearls Publishing; 2020.

3. Davis AR, Castaño PM. Oral contraceptives and libido in women. *Annu Rev Sex Res*. 2004;15:297–320.

4. Casado-Espada NM, de Alarcón R, de la Iglesia-Larrad JI, Bote-Bonaechea B, Montejo ÁL. Hormonal contraceptives, female sexual dysfunction, and managing strategies: A review. *J Clin Med*. 2019;8(6):908.

5. Almagor M, Ben-Porath YS. Mood changes during the menstrual cycle and their relation to the use of oral contraceptive. *J Psychosom Res*. 1991;35(6):721–728.

6. Oinonen KA, Mazmanian D. To what extent do oral contraceptives influence mood and affect? *J Affect Disord*. 2002;70(3):229–240.

7. Smith NK, Jozkowski KN, Sanders SA. Hormonal contraception and female pain, orgasm and sexual pleasure. *J Sex Med*. 2014;11(2):462–470.

8. Both S, Lew-Starowicz M, Luria M, et al. Hormonal contraception and female sexuality: Position statements from the european society of sexual medicine (ESSM). *J Sex Med*. 2019;16(11):1681–1695.

9. Casey PM, MacLaughlin KL, Faubion SS. Impact of contraception on female sexual function. *J Womens Health (Larchmt)*. 2017;26(3):207–213.

10. Wallwiener CW, Wallwiener LM, Seeger H, Mück AO, Bitzer J, Wallwiener M. Prevalence of sexual dysfunction and impact of contraception in female German medical students. *J Sex Med.* 2010;7(6):2139–2148.

11. Heidari M, Ghodusi M, Rezaei P, Kabirian Abyaneh S, Sureshjani EH, Sheikhi RA. Sexual function and factors affecting menopause: A systematic review. *J Menopausal Med.* 2019;25(1):15–27.

12. Bloch A. Self-awareness during the menopause. *Maturitas.* 2002;41(1):61–68.

13. Jonusiene G, Zilaitiene B, Adomaitiene V, Aniuliene R, Bancroft J. Sexual function, mood and menopause symptoms in Lithuanian postmenopausal women. *Climacteric.* 2013;16(1):185–193.

14. Masliza W, Daud W, Yazid Bajuri M, et al. Sexual dysfunction among postmenopausal women. *Clin Ter.* 2014;165(2):83–89.

15. Blümel JE, Chedraui P, Baron G, et al. Sexual dysfunction in middle-aged women: A multicenter Latin American study using the Female Sexual Function Index. *Menopause.* 2009;16(6):1139–1148.

16. National Cancer Institute. Menopausal hormone therapy and cancer. https://www.cancer.gov/about-cancer/causes-prevention/risk/hormones/mht-fact-sheet. Published 2020. Accessed August 31, 2020.

17. Castelo-Branco C, Blumel JE, Araya H, et al. Prevalence of sexual dysfunction in a cohort of middle-aged women: Influences of menopause and hormone replacement therapy. *J Obstetr Gynaecol* 2003;23(4):426–430.

18. Ziaei S, Moghasemi M, Faghihzadeh S. Comparative effects of conventional hormone replacement therapy and tibolone on climacteric symptoms and sexual dysfunction in postmenopausal women. *Climacteric.* 2010;13(2):147–156.

19. Del Río JP, Molina S, Hidalgo-Lanussa O, Garcia-Segura LM, Barreto GE. Tibolone as hormonal therapy and neuroprotective agent. *Trends Endocrinol Metab.* 2020;31(10):742–759.

20. Anagnostis P, Bitzer J, Cano A, et al. Menopause symptom management in women with dyslipidemias: An EMAS clinical guide. *Maturitas.* 2020;135:82–88.

21. Austad SN. Menopause: An evolutionary perspective. *Exp Gerontol.* 1994;29(3-4):255–263.

22. Walker ML, Herndon JG. Menopause in nonhuman primates? *Biol Reprod.* 2008;79(3):398–406.

23. University of Cambridge. Menopause is an adaptation to minimize reproductive competition between females in a family. *ScienceDaily.* https://www.sciencedaily.com/releases/2008/03/080331172519.htm. Published 2008. Accessed August 31, 2020.

8 Pregnancy and Lactation

8.1 INTRODUCTION

Drugs taken during pregnancy can sometimes cross the placenta (Fig. 8.1) and enter the body of the fetus. For that reason, many physicians are reluctant to prescribe to a pregnant person even familiar drugs that are otherwise generally considered safe. Drugs taken by nursing mothers (Fig. 8.2) have the potential to reach the baby. In fact, most drugs will be secreted in breast milk, although sometimes only in trace amounts. This may pose a problem for those who become pregnant (Fig. 8.3), when they have medical conditions requiring medical therapy. Also, pregnant women can develop new conditions or need specific treatments, such as therapy for migraine headaches (Fig. 8.4). In such cases, physicians must consider the risks of offering no drug treatments against the risks of offering treatment. Guidance for prescribing to pregnant and lactating patients can be scarce, as many clinical trials routinely exclude them. The United States Food and Drug Administration (FDA) has strict guidance regarding how drugs may be labeled with respect to their use in people who are pregnant. Category A drugs show no risk of fetal abnormalities and Category D drugs have demonstrated a risk for such damages; Category X drugs are those in which well-controlled or observational studies have demonstrated evidence of fetal abnormalities—these drugs must be considered contraindicated for those who are or might become pregnant.[1]

Not all drugs must be discontinued during pregnancy and lactation; there are numerous facets to the issue of pharmacological therapy at this stage of life. First, a drug may affect the developing embryo or fetus, but it may also affect the mother. Second, the mother's physiology changes markedly as pregnancy advances, with significant weight gain the most obvious of these changes. These changes can alter how drugs are metabolized. And finally, much depends on the drugs taken, the dose and frequency, the maternal and fetal status, how the blood is distributed in the maternal bloodstream, and whether the drug can cross the placenta. Drugs can also be transferred to the mother's milk during lactation and be ingested by the baby.

8.2 DRUGS IN PREGNANCY

8.2.1 Teratogenic Effects

During blastogenesis (Fig. 8.5), in the first few weeks after fertilization, the embryo can be affected by drugs in a most extreme way: they tend to either destroy the embryo or have no effect at all. The embryo is particularly vulnerable to birth defects during the period of organogenesis, occurring between the third and eighth weeks of pregnancy.[1] After nine weeks (Fig. 8.6), the fetus matures and is less susceptible to birth defects.[1]

Some drugs are known to be teratogenic and can harm a developing fetus, sometimes in profound ways. In the 1960s, thalidomide (Fig. 8.7) was widely used by pregnant women—with disastrous consequences, as this agent, which today is used in chemotherapy, can result in severe birth defects.[2] Today, the acne drug isotretinoin is known as a powerful teratogenic, carrying a risk of 20%–35% for congenital defects in those exposed to it in the uterus.[3] As a result, in people who could become pregnant, isotretinoin therapy is only initiated after a negative pregnancy test and counseling about the risks of the

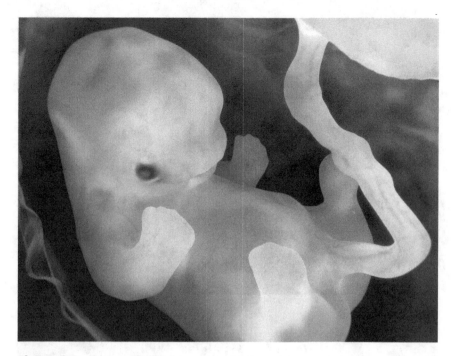

Figure 8.1 Fetus and placenta.

Figure 8.2 Nursing baby.

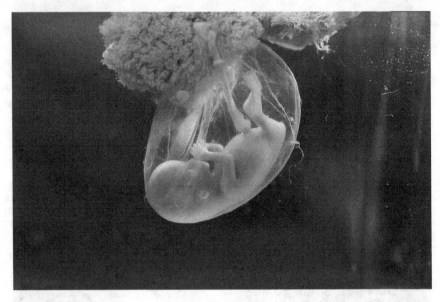

Figure 8.3 Fetus.

Migraine

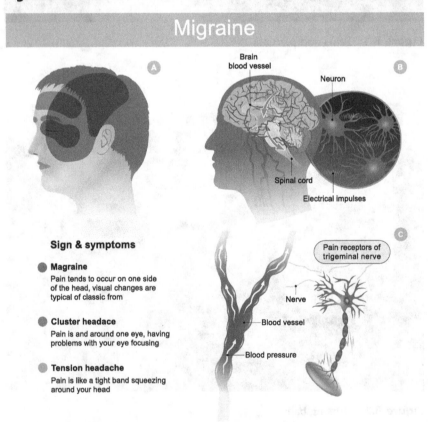

Sign & symptoms

● **Magraine**
Pain tends to occur on one side of the head, visual changes are typical of classic from

● **Cluster headace**
Pain is and around one eye, having problems with your eye focusing

● **Tension headache**
Pain is like a tight band squeezing around your head

Figure 8.4 Migraine headache.

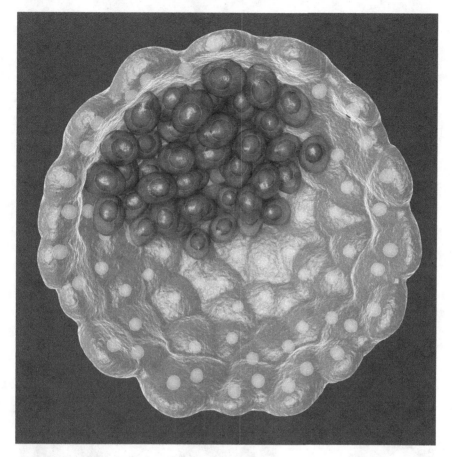

Figure 8.5 Blastogenesis.

drug. It no longer thought to pose a risk after five elimination half-lives, so pregnancy should be avoided for at least one month after drug cessation. But this is now recommended for extension to three months, due to variability in the drug's half-life.[4-6] The iPledge program in the United States asks that those who are prescribed isotretinoin sign a document stating that they are aware of the drug's risks. This program is required for all people who take isotretinoin, including men and women who are past childbearing potential.

While drug-induced birth defects can occur, it is estimated that of all birth defects, only about 2%–3% are the result of drugs taken during the pregnancy.[1]

8.2.2 Maternal Changes in Pregnancy

Changes during the course of pregnancy can affect how drugs are metabolized (Fig. 8.8). During pregnancy, blood plasma volume increases by 30%–50%, which may result in lower concentrations of circulating agents or drug accumulation. As body fat increases during pregnancy, distribution of lipophilic drugs may be altered. Plasma albumin decreases during pregnancy, which means that highly protein-bound drugs have an increased distribution. By contrast, unbound drugs are more likely to be excreted more rapidly. Gastric emptying becomes more and more delayed in the third trimester of pregnancy, which can delay a drug's onset of action.

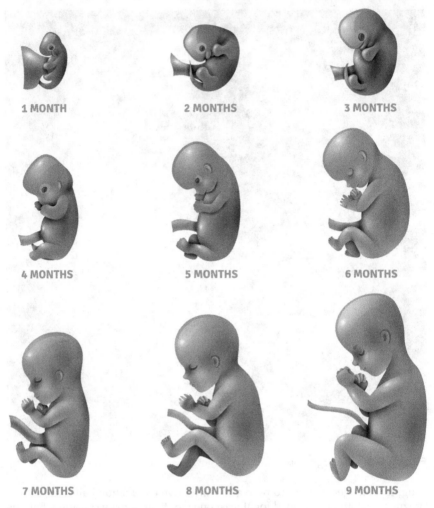

Figure 8.6 Fetal development by month.

Many pregnant women take prenatal supplements, including iron, which can serve to bind or inactivate certain drugs. Estrogen and progesterone can affect the activity of liver enzymes, which in turn can cause some drugs to accumulate or others to be eliminated more rapidly.[7]

8.2.3 Crossing the Placenta

In order for a drug to affect the fetus, it must travel from maternal circulation via the placenta into the fetal circulation. About 50 years ago, it was believed that the placenta was an impenetrable protector, but this is not the case. There are some drugs (free, unbound drugs) that can cross the placenta. Since the fetal pH tends to be slightly more acidic than the maternal pH, weak bases have an easier time crossing the placenta.[8] Drugs with lower molecular weights (500–1,000 g/mol) are facilitated in crossing the placenta, whereas drugs with molecular weights above 1,000 g/mol do not seem to be able to get across it.[9] As the pregnancy advances, placental transfer becomes easier,

92

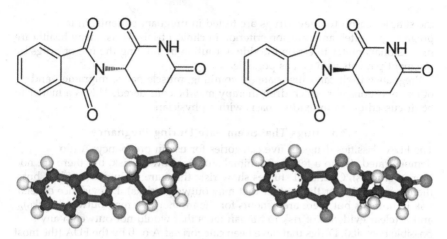

Figure 8.7 Chemical structure of thalidomide.

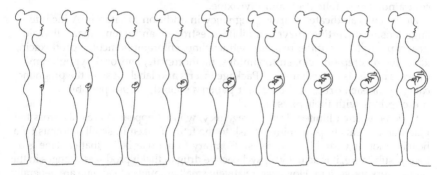

Figure 8.8 Changes during pregnancy can affect the actions of drugs.

because there is more and more maternal blood movement into the placenta, and the surface area of the placenta increases.

In addition to drugs traveling over the placenta to the fetal bloodstream, drugs can constrict the placental vasculature and thus limit the amount of oxygen and nutrients delivered to the fetus. Such situations may result in an underweight or underdeveloped baby.

8.2.4 Should a Pregnant Person Take Drugs?

Many women who take prescription, over-the-counter, or recreational drugs may take them before realizing that they are pregnant. After pregnancy is confirmed, some women still take over-the-counter or recreational drugs, which can be problematic for the mother, child, and obstetrics team. However, there are situations in which a mother needs to take medications or the cessation of pharmacological therapy would actually represent potential maternal or fetal harm. In such cases, physicians are often called upon to help reach decisions about ongoing drug therapies and whether or not new drug treatments should be started to address medical problems that arise during pregnancy. The great problem is that, for obvious reasons, few drugs are thoroughly tested in the pregnant population, resulting in limited guidance for prescribing physicians. Many package inserts do not give guidance about drugs during pregnancy for the simple reason that few drugs are tested in pregnant patients. In fact, pregnancy is often an exclusion criterion in clinical trials. Thus, many healthcare professionals prefer to err on the side of caution by limiting the use of drugs by pregnant patients as much as possible.

Pregnancy itself can cause nausea, vomiting, muscle pains, migraines, and other conditions, for which drug therapy may be considered. This is a matter to be discussed on an individual basis with a physician.

8.2.5 Drugs That Seem Safe During Pregnancy

The FDA classifies drugs in five categories for use in pregnancy: A (no demonstrated risk to a fetus), B (animal studies show no risk, but there are no human studies), C (animal studies show risk, there are no human studies, but there are benefits for the mother that may outweigh risks), D (evidence of fetal risk in humans but there are benefits for the mother that may outweigh risks), and X (clear evidence of risk to human fetus that would not outweigh any possible benefits). Drugs that have been categorized A or B by the FDA (the most benign drugs) include acetaminophen, doxylamine, meclizine, cyclizine, dimenhydrinate, many antibiotics, vitamins B, C, D, and E (vitamin A is contraindicated), folic acid, and thyroxine.

Drugs contraindicated during pregnancy in addition to vitamin A include angiotensin-converting enzyme inhibitors, estrogen, androgens, many thyroid drugs, phenytoin, antidepressants, tetracycline, chloramphenicol, ciprofloxacin, kanamycin, streptomycin, sulfonamides, methotrexate, and oral hypoglycemic drugs. This list is not exhaustive. Package inserts and labeling state the pregnancy category of each drug, and pregnant patients should discuss possible prescriptions with their physician.

Caffeine is often limited during pregnancy; while it appears to readily cross the placenta, it seems to pose minimal risk to the fetus if taken in small amounts by a healthy mother with a healthy fetus.[1] Contrary to popular belief that caffeine is a benign substance, it is actually a psychoactive drug in that it produces changes in the central nervous system. However, relatively small amounts of caffeine are generally not risky, and are considered socially acceptable as well. Cigarette smoking and alcohol consumption are far riskier, and should be avoided during pregnancy and lactation. Illicit drugs can be extremely dangerous. Cocaine, for example, readily crosses the placenta, constricts placental blood flow, and can result in serious complications during the pregnancy for the mother and preterm delivery. Cocaine use during pregnancy increases the chances of miscarriage and placental detachment.[1] Mothers who take potentially addicting drugs during their pregnancy may give birth to babies who are dependent on those drugs and will endure withdrawal symptoms after birth. In addition, opioids can cause lower-than-normal birth weight, which can pose significant risk to the newborn.

8.3 DRUGS AND LACTATION

Mother's, or surrogate's, milk is a complex fluid, with numerous antibodies, enzymes, hormones, and nutrients that confer notable morbidity and mortality benefits to a baby.[10] Milk synthesis begins in the pregnant woman's body at about 10–22 weeks of gestation. Milk availability occurs at about 48 hours after delivery, but this is mainly colostrum. When maternal progesterone levels decrease, milk production becomes more plentiful; this usually happens around four days after delivery. However, mothers who deliver preterm babies may have a longer period before milk production is plentiful. The pituitary gland secretes prolactin that causes milk to be produced in response to breast stimulation; oxytocin is produced from the pituitary to aid in milk production. When the mother is under stress or in great pain, oxytocin production may be reduced, and this can limit the milk supply.[10]

During lactation, it is possible for substances from the mother's body to be transferred to the breast milk. In fact, nearly all drugs that are taken by a lactating mother will appear in her breast milk, albeit sometimes in only trace amounts.[11] Surprisingly, most drugs are not of serious concern for nursing mothers; the situation is further improved by the fact that many lactating women exercise caution about pharmaceuticals during nursing and take few or no drugs.[12] If a nursing mother is prescribed a drug, she should discuss it with her physician. In general, drugs contraindicated during lactation are amiodarone (a heart drug with a very long half-life), iodine, lithium, oral retinoids, gold salts (for nephritis), and certain cancer drugs (antineoplastics and radiopharmaceuticals).[11] Insulin and heparin, for example, are not contraindicated, as their molecules are too large to enter breast milk.[11]

Many factors can influence drug concentration in mother's, or surrogate's, milk. For instance, the drug's distribution in the mother's tissue and the time course of taking the drug may influence concentration, as may protein binding. Drugs that have high protein binding are less likely to be transferred to the baby, or will be transferred only in minute amounts. Drugs with lower protein binding are more likely to be transferred.[11]

Codeine can be extremely dangerous for nursing mothers and their babies. A subset of the population (about 10% of North Americans but up to 30% of North Africans) has a genetic polymorphism of the cytochrome P450 2D6 enzyme that causes them to rapidly metabolize codeine. These so-called ultra-rapid metabolizers can produce large quantities of morphine metabolites, which, if transferred to a baby during nursing, could cause codeine toxicity, respiratory depression, and even death.[11] Codeine should be avoided by lactating mothers; if analgesics are required, alternative agents or nonpharmacological treatments should be selected.

Some drugs can inhibit milk production but may not be dangerous to the baby. These include antihistamines, birth control pills with estrogen, decongestants (pseudoephedrine), and fertility drugs (such as clomiphene). Even modest alcohol consumption can reduce breast milk production.

Package inserts with drugs often state that the drugs should be avoided during lactation. This may be done more for regulatory than medical reasons. Pharmaceutical manufacturers cannot state with certainty on the package insert that the drug is safe without sufficient and extensive testing, and clinical testing of nursing mothers is rare. For that reason, it may be the case that a drug is not recommended during lactation but is actually safe.

In some cases, new mothers must make a decision between taking a drug they need and nursing their baby. Such decisions can only be made on an individual basis and should consider all the factors: the safety of the drug, the dangers of delaying or avoiding treatment, the availability of safer alternatives, and how much breast milk the child consumes (neonates consume almost entirely breast milk, but as babies get older, other foods are introduced).

8.4 CONCLUSION

A database for healthcare professionals about drugs during lactation is available at LactMed, a free US-based service from the National Center for Biotechnology Information with extensive links to the PubMed database. The Breastfeeding Network, based in the United Kingdom, also provides extensive information about drugs during lactation. Other such resources are available online as well. In using such resources, it is important to realize that many drugs are not tested in people who are lactating, and an absence of evidence that a drug is harmful is not the same thing as the drug being safe. Overall, the fewer drugs a lactating mother, or surrogate, takes, the better. When drugs are necessary, the prescription should be discussed with the physician.

REFERENCES

1. Sachdeva P, Patel BG, Patel BK. Drug use in pregnancy: A point to ponder! *Indian J Pharm Sci.* 2009;71(1):1–7.

2. Vargesson N. Thalidomide-induced teratogenesis: History and mechanisms. *Birth Defects Res C Embryo Today.* 2015;105(2):140–156.

3. Choi JS, Koren G, Nulman I. Pregnancy and isotretinoin therapy. *Canad Med Assoc J.* 2013;185(5):411–413.

4. Wiegand UW, Chou RC. Pharmacokinetics of oral isotretinoin. *J Am Acad Dermatol.* 1998;39(2 Pt 3):S8–12.

5. Nulman I, Berkovitch M, Klein J, et al. Steady-state pharmacokinetics of isotretinoin and its 4-oxo metabolite: Implications for fetal safety. *J Clin Pharmacol.* 1998;38(10):926–930.

6. Lee SM, Kim HM, Lee JS, et al. A case of suspected isotretinoin-induced malformation in a baby of a mother who became pregnant one month after discontinuation of the drug. *Yonsei Med J.* 2009;50(3):445–447.

7. Hansen WF, Yankowitz J. Pharmacologic therapy for medical disorders during pregnancy. *Clin Obstet Gynecol.* 2002;45(1):136–152.

8. Loebstein R, Lalkin A, Koren G. Pharmacokinetic changes during pregnancy and their clinical relevance. *Clin Pharmacokinet.* 1997;33(5): 328–343.

9. Kraemer K. Placental transfer of drugs. *Neonatal Netw.* 1997;16(2):65–67.

10. McGuire TM. Drugs affecting milk supply during lactation. *Aust Prescr.* 2018;41(1):7–9.

11. Hotham N, Hotham E. Drugs in breastfeeding. *Aust Prescr.* 2015;38(5): 156–159.

12. Ilett KF, Kristensen JH. Drug use and breastfeeding. *Expert Opin Drug Saf.* 2005;4(4):745–768.

9 Recreational/Illicit Drugs and Sex

9.1 INTRODUCTION

The extent of use of drugs for recreational purposes is high; in 2017, about 11% of all people over age 12 in the United States had used an illicit drug within the past month.[1] Changing laws and attitudes about marijuana consumption have added to that number, in particular in areas where recreational marijuana use is legal and even promoted. Along with alcohol, drugs are frequently used to facilitate sexual contact and enhance sexual pleasure.

9.2 MARIJUANA

Marijuana (Fig. 9.1) is a ubiquitous drug in many parts of the world, where it enjoys general public tolerance if not legal acceptance. Marijuana is associated with feelings of well-being that make it a popular drug for social gatherings and romantic partners. The use of marijuana among college students has been associated with an openness to new experiences, which may lead to sexual adventures.[2] This may not be a causal relationship, but could rather reflect the fact that the same things that make a young person eager to experiment with marijuana may make that person open to sexual activity.

Despite its anecdotal reputation as a feel-good drug for romantic partners, there is a lack of strong evidence that marijuana can increase desire, encourage arousal, improve performance, or lead to greater sexual satisfaction. As early as in 1979, a psychology journal reported that marijuana enhanced sexual activity in a study of 84 graduate students.[3] A subsequent survey of 216 men and women who said they regularly had sex under the influence of marijuana reported that 39% of respondents said the sex was better, 16% said it was better in some ways but worse in others, 24% said marijuana only sometimes made sex better, and 5% said sex was worse with marijuana.[4] Most respondents found that marijuana increased their sexual desire (59%), and 74% said it improved their satisfaction with sex, with 66% saying it made orgasms more intense. About 70% of respondents said that marijuana helped them relax, which facilitated the sexual encounter.[4]

The strains of marijuana available today have substantially higher content of delta-9-tetrahydrocannabinol (THC; Fig. 9.2) than the marijuana of the 1960s and 1970s. THC is a psychoactive substance that produces the marijuana high and affects the neurological and endocrine systems. There may be sex-specific differences in how THC affects the brain and subsequent actions, but these remain to be elucidated.[5] For example, it has been reported that men may derive greater analgesic benefit from marijuana than women,[6] and women may experience more pronounced effects than men in terms of attention and cognition.[7]

Contrary to popular opinion, marijuana does not necessarily acutely increase risk-taking behaviors in sexual acivity.[8] However, a study from China based on data sets from US National Longitudinal Study of Adolescent to Adult Health found that marijuana use was associated with sex with multiple partners, which in some cases might be considered risk-taking behavior.[9] A study from Spain reported that marijuana use increased the likelihood of unsafe sex, defined as sex without a condom.[10] Daily cannabis use was reported in a survey ($n = 8,650$) to be associated with sexually transmitted diseases in women but not men.[11] In addition, frequency of marijuana use has been associated with greater frequency

Figure 9.1 Marijuana leaves.

Figure 9.2 Chemical structure of tetrahydrocannabinol.

of sexual contact and more sexual satisfaction in women, but this same dose-dependent relationship with sexual contact was not true for men.[12]

There is some equivocal and some contradictory evidence in the literature that marijuana may be associated with erectile dysfunction.[13] However, long-term regular use of marijuana, particularly with high THC content, may lead to a dysfunction of the smooth muscle of the penis, which might result in erectile dysfunction.[11] Studies about the adverse effects of marijuana on male reproductive function have not been conclusive.[14]

9.3 STIMULANTS

Stimulant drugs include gamma hydroxybutyrate, mephedrone, amphetamines, and methamphetamine (including crystal meth). Methamphetamine is the N-methyl derivative of amphetamine; although once used medicinally to treat narcolepsy, it was removed from the market because of its potential for abuse.[15] It can produce euphoric, empathogenic, and even hallucinogenic effects.

In a survey of 1,159 men who used illicit methamphetamine (but no other illicit drugs), half said it did not affect their sexual performance, while the other half reported that it caused some sexual effects.[16] These effects ranged from greater orgasmic intensity and prolonged ejaculation latency to reduced erectile rigidity and lower sexual satisfaction. Compared to 211 age-matched control subjects, the methamphetamine users reported lower scores in terms of erectile function, orgasmic satisfaction, and sexual desire, and higher rates of erectile dysfunction (29% vs. 12%).[16]

Gamma hydroxybutyric acid (GHB) and its congeners are potent depressors of the central nervous system and are frequently reported in chemsex encounters (see later). In fact, GHB is often taken and described in the context of sexual activities. Its effect is usually felt 10–30 minutes after ingestion, and can last up to six hours. In a survey of 60 recreational GHB users, 26% said it increased sexual arousal and made them more attracted to their partner, and 35% reported that they engaged in sexual intercourse with strangers while taking this drug.[17]

A particular risk for GHB use is that blackouts can occur; in that same survey of 60 recreational users of GHB, 25% had experienced a GHB-related blackout.[17] During loss of consciousness, GHB users are not only at risk for morbidity and mortality, but also of rape and sexual assault. Furthermore, GHB dependence, overdose toxicity, and withdrawal symptoms have been reported.[18] Very little information is available on adulterated GHB available on the street market, although such tainted products are presumed to exist.[15]

9.4 MDMA (3,4-METHYLENEDIOXYMETHAMPHETAMINE)

3,4-Methylenedioxymethamphetamine (MDMA), sometimes called ecstasy, X, or Molly, may be used to enhance feelings of intimacy between sexual partners, although it is not a typical "sex drug." Technically, MDMA is a stimulant and a popular "party drug." It has been reported that some men use MDMA in sexual situations to enhance their sexual pleasure, whereas some women use it to help manage psychological or physical discomfort with sexual activity.[19] MDMA may result in increased production of cortisol, prolactin, and oxytocin.[20] In a survey of 98 MDMA users over the course of six months, most reported that the drug increased feelings of closeness and intimacy with their partners, leading to sexual arousal, but it did not specifically increase the desire for sexual penetration.[21]

MDMA use also increases risky sexual behaviors.[21] It is sometimes used in combination with alcohol or marijuana, whose effects it is thought to potentiate. MDMA can result in dependence, and long-term use has been associated with increased rates of depression and impaired memory.

9.5 COCAINE

Cocaine and its by-product, crack (small chunks called rocks that are cheaper than cocaine powder and may be smoked or inhaled), have paradoxical effects on sex.

Cocaine is a stimulant (inhibits the neuronal reuptake of norepinephrine) with analgesic effects that carries a high risk for dependence with continuing use. In fact, about 5% of those who try cocaine will develop cocaine use disorder within the first year.[22] Regular users of cocaine describe the psychostimulant drug as producing effects of euphoria, high energy, stimulation, confidence, exhilaration, and heightened sensory perceptions. It enjoys a popular reputation of promoting sexual interest, and many believe that it improves sexual performance and enhances sexual pleasure. Paradoxically, regular or long-term cocaine use can cause sexual dysfunction in the form of erectile dysfunction in men and anorgasmia in women. Likewise, cocaine is believed by many users to prolong sexual activity and increase orgasmic intensity, but over time it may result in difficulties in achieving orgasm, if not outright anorgasmia.

Cocaine use and cocaine use disorder are associated with risk-taking behaviors and impulsivity, but it is not clear whether those who use cocaine are inherently impulsive or it is the cocaine that makes them impulsive. Impulsivity in this context may be defined as a tendency to make fast and unplanned decisions in response to external or internal stimuli, with little to no reflection on the consequences or results of such actions. Impulsive sexual behaviors including having sex with strangers, having sex with multiple partners, having sex in public places, and having unsafe (unprotected) sex. In a survey of 350 cocaine users in rehabilitation in Brazil, 40% said they never used a condom during sexual intercourse, and the rest of the espondents said their condom use was rare or occasional.[23] Thus, the incidence of HIV infection is relatively high (range 4.0%–10.2%) among those who regularly use cocaine or crack cocaine, likely due to unsafe sex.[23] While statistics are not available, many people with cocaine use disorder and few financial resources resort to occasional or even full-time sex work to support their habit. The incidence of violence and antisocial behavior (fighting, threatening others, theft) is also higher among regular cocaine users, although sexual intercourse may be less frequent than among nonusers.[23]

About 25% of regular cocaine users report occasional episodes of bingeing, or ingesting large amounts of cocaine without pause over a limited period of time.[24] Binges are intense and usually end only when the drugs and money run out. Cocaine binges have been significantly associated with having sex under the influence of cocaine.[24] A typical binge lasts about 3.7 days and for those who use crack, it involves an average of about 40 "rocks." In a survey of 303 HIV-positive Black cocaine users, 72% had sex during their last cocaine binge, with an average of 3.1 partners.[25]

Substance-use disorder is often comorbid with mental health disorders, and there is a strong link between borderline personality disorder and specifically cocaine or crack cocaine dependence. Borderline personality disorder is characterized by pronounced risk-taking and self-destructive behaviors.[26] About 19% of people with cocaine dependence have clinical borderline personality disorder, although this comorbidity is not well studied.[22] In fact, cocaine use disorder is a predictor of borderline personality disorder, with an odds ratio of 2.06.[22]

9.6 CHEMSEX

'Chemsex' refers to the use of multiple drugs, both psychoactive and not, with or without alcohol, in a recreational party setting aimed at fostering or enhancing sexual activity. Heterosexual chemsex is clearly common, but the term currently is most frequently discussed in terms of sex among men who have sex with other men.[27] Chemsex among women who have sex with other women is currently less commonly reported and is not well studied.

The drugs of a chemsex cocktail are taken for a variety of reasons: for example, to prolong sexual activity, to allow for sex with multiple partners on the same evening, to improve sexual performance, to enhance sexual pleasure, and to facilitate sexual encounters. Chemsex involves polysubstance use and is mainly associated with drugs like GHB, methamphetamine, drugs to combat erectile dysfunction (sildenafil, tadalfil, vardenafil), and alkyl nitrites (amyl nitrites and butyl nitrites, also known as "poppers"), although other drugs may also be used.[15] Alcohol is not a main substance of a chemsex combination, but it may be provided at chemsex parties. The most frequently reported drug combination reported in the literature is an interesting mix of methamphetamine, GHB, and an erectile dysfunction drug.[15] The GHB is a depressant of the central nervous system, while the methamphetamine acts as a stimulant; taken together, they might be thought to elicit feelings of sexual arousal, excitement, and enthusiasm. The use of amyl nitrites decreases pain perception, relaxes muscles, and may produce mild hallucinations, the latter of which can be stimulated by music and lighting effects, making these popular club drugs.[28] Erectile dysfunction drugs promote reliable and prolonged erections, although alkyl nitrites (one type of poppers) can interact with these drugs, even provoking potentially life-threatening side effects such as stroke or heart attack.[29] All of the popper drugs can have side effects, ranging from headache to retinal toxicity.[15] Methamphetamine can lead to dependence and other adverse effects.[15]

Chemsex is an international phenomenon, and not limited to urban milieus. There is another important public health consideration with chemsex beyond unsafe sex and the potential spread of HIV and other sexually transmitted diseases, namely that some participants are introduced to specific and highly addictive illicit drugs, such as methamphetamine, through chemsex parties.[30,31] Although chemsex is decades old, large longitudinal studies are lacking that might provide evidence as to the relationship between HIV infections or substance-use disorders and involvement with chemsex events.

9.7 PEOPLE WITH SUBSTANCE-USE DISORDERS

While there is an abundance of medical literature about specific illicit drugs and sexual activity, there is a paucity of literature on sexual behaviors among people with active substance-use disorders. A substance-use disorder may be defined in different ways, but for purposes here it generally involves the compulsion experienced by an individual to ingest a substance, often having intense cravings for it, even if that person realizes it is harmful. In a survey of 180 people in treatment for substance-use disorders, participants were grouped by their substance of choice: alcohol, stimulants (various amphetamines), sedatives (opioids, benzodiazepines), and GHB.[32] All groups reported that their substance use changed their attitudes and behaviors with regard to sexual activity, with about half reporting that the substance improved sex for them. About a quarter of respondents said that their attitudes and feelings about sex had become closely intertwined with the substance use.[32] In other words, people who had sex under the influence of heroin soon came to think of sex as something they did while taking heroin; the substance soon eclipsed the sexual activity.

Sexual addiction (according to self-reported instruments) has been found to be more frequent among individuals with a substance-use disorder and some form of polysubstance use (odds ratio, 2.72; 95% confidence interval, 1.1–6.7) than among those whose substance-use disorder involves only one substance (odds ratio, 0.3; 95% confidence interval, 1.15–0.91).[33] In that same study, among those who used crack or cocaine, either alone or with other illicit substances, greater drug consumption correlated to a higher degree of sexual addiction.

9.8 CONCLUSION

Many people use recreational drugs to enhance or facilitate sex. Because some of these drugs are illegal and in many cases these activities would be illegal or subject to criminal prosecution, researchers can have difficulty studying this phenomenon. In addition, 'street drugs' are of variable quality, may be adulterated, and are often taken opportunistically. Thus, there is much that remains to be elucidated about illicit drugs and sexual behaviors, and in some cases these drugs enjoy a popular street reputation that may not entirely line up with their actual effects.

REFERENCES

1. Centers for Disease Control and Prevention. Illicit drug use. https://www. cdc.gov/nchs/fastats/drug-use-illicit.htm. Published 2019. Accessed December 9, 2020.

2. Phillips KT, Phillips MM, Duck KD. Factors associated with marijuana use and problems among college students in Colorado. *Subst Use Misuse.* 2018;53(3):477–483.

3. Dawley HH, Jr., Winstead DK, Baxter AS, Gay JR. An attitude survey of the effects of marijuana on sexual enjoyment. *J Clin Psychol.* 1979;35(1):212–217.

4. Wiebe E, Just A. How cannabis alters sexual experience: A survey of men and women. *J Sex Med.* 2019;16(11):1758–1762.

5. Ketcherside A, Baine J, Filbey F. Sex effects of marijuana on brain structure and function. *Curr Addict Rep.* 2016;3:323–331.

6. Cooper ZD, Haney M. Sex-dependent effects of cannabis-induced analgesia. *Drug Alcohol Depend.* 2016;167:112–120.

7. Anderson BM, Rizzo M, Block RI, Pearlson GD, O'Leary DS. Sex, drugs, and cognition: Effects of marijuana. *J Psychoactive Drugs.* 2010;42(4):413–424.

8. Skalski LM, Gunn RL, Caswell A, Maisto S, Metrik J. Sex-related marijuana expectancies as predictors of sexual risk behavior following smoked marijuana challenge. *Exp Clin Psychopharmacol.* 2017;25(5):402–411.

9. Zhang X, Wu LT. Marijuana use and sex with multiple partners among lesbian, gay and bisexual youth: Results from a national sample. *BMC Public Health.* 2017;17(1):19.

10. Moure-Rodríguez L, Doallo S, Juan-Salvadores P, Corral M, Cadaveira F, Caamaño-Isorna F. Heavy episodic drinking, cannabis use and unsafe sex among university students. *Gac Sanit.* 2016;30(6):438–443.

11. Smith AM, Ferris JA, Simpson JM, Shelley J, Pitts MK, Richters J. Cannabis use and sexual health. *J Sex Med.* 2010;7(2 Pt 1):787–793.

12. Kasman AM, Bhambhvani HP, Wilson-King G, Eisenberg ML. Assessment of the association of cannabis on female sexual function with the female sexual function index. *Sex Med.* 2020;8(4):699–708.

13. Shamloul R, Bella AJ. Impact of cannabis use on male sexual health. *J Sex Med.* 2011;8(4):971–975.

14. Hsiao P, Clavijo RI. Adverse effects of cannabis on male reproduction. *Eur Urol Focus.* 2018;4(3):324–328.

15. Giorgetti R, Tagliabracci A, Schifano F, Zaami S, Marinelli E, Busardò FP. When "chems" meet sex: A rising phenomenon called "chemsex". *Curr Neuropharmacol.* 2017;15(5):762–770.

16. Chou NH, Huang YJ, Jiann BP. The impact of illicit use of amphetamine on male sexual functions. *J Sex Med.* 2015;12(8):1694–1702.

17. Kapitány-Fövény M, Mervó B, Corazza O, et al. Enhancing sexual desire and experience: An investigation of the sexual correlates of gamma-hydroxybutyrate (GHB) use. *Human Psychopharmacol.* 2015;30(4):276–284.

18. Brennan R, Van Hout MC. Gamma-hydroxybutyrate (GHB): A scoping review of pharmacology, toxicology, motives for use, and user groups. *J Psychoactive Drugs.* 2014;46(3):243–251.

19. Kostick KM, Schensul JJ. The role of ecstasy (MDMA) in managing intimacy and conflict in stable relationships. *Cult Health Sex.* 2018;20(10):1071–1086.

20. Dolder PC, Müller F, Schmid Y, Borgwardt SJ, Liechti ME. Direct comparison of the acute subjective, emotional, autonomic, and endocrine effects of MDMA, methylphenidate, and modafinil in healthy subjects. *Psychopharmacology.* 2018;235(2):467–479.

21. McElrath K. MDMA and sexual behavior: Ecstasy users' perceptions about sexuality and sexual risk. *Subst Use Misuse.* 2005;40(9-10):1461–1477.

22. Balducci T, González-Olvera JJ, Angeles-Valdez D, Espinoza-Luna I, Garza-Villarreal EA. Borderline personality disorder with cocaine dependence: Impulsivity, emotional dysregulation and amygdala functional connectivity. *Front Psychiatry.* 2018;9:328.

23. Carvalho HB, Seibel SD. Crack cocaine use and its relationship with violence and HIV. *Clinics (Sao Paulo).* 2009;64(9):857–866.

24. Roy É, Arruda N, Jutras-Aswad D, et al. Examining the link between cocaine binging and individual, social and behavioral factors among street-based cocaine users. *Addict Behav.* 2017;68:66–72.

25. Harzke AJ, Williams ML, Bowen AM. Binge use of crack cocaine and sexual risk behaviors among African-American, HIV-positive users. *AIDS Behav.* 2009;13(6):1106–1118.

26. Tull MT, Gratz KL, Weiss NH. Exploring associations between borderline personality disorder, crack/cocaine dependence, gender, and risky sexual behavior among substance-dependent inpatients. *Personal Disorders: Theory, Res Treatm.* 2011;2(3):209–219.

27. Pirani F, Lo Faro AF, Tini A. Is the issue of chemsex changing? *Clin Ter.* 2019;170(5):e337–e338.

28. Lowry TP. Psychosexual aspects of the volatile nitrites. *J Psychoactive Drugs.* 1982;14(1-2):77–79.

29. Romanelli F, Smith KM. Recreational use of sildenafil by HIV-positive and - negative homosexual/bisexual males. *Ann Pharmacother.* 2004;38(6): 1024–1030.

30. Sewell J, Cambiano V, Speakman A, et al. Changes in chemsex and sexual behaviour over time, among a cohort of MSM in London and Brighton: Findings from the AURAH2 study. *Int J Drug Policy.* 2019;68:54–61.

31. Bourne A, Reid D, Hickson F, Torres-Rueda S, Steinberg P, Weatherburn P. "Chemsex" and harm reduction need among gay men in South London. *Int J Drug Policy.* 2015;26(12):1171–1176.

32. Bosma-Bleeker MH, Blaauw E. Substance use disorders and sexual behavior: The effects of alcohol and drugs on patients' sexual thoughts, feelings and behavior. *Addict Behav.* 2018;87:231–237.

33. Antonio N, Diehl A, Niel M, et al. Sexual addiction in drug addicts: The impact of drug of choice and poly-addiction. *Rev Assoc Med Bras (1992).* 2017;63(5):414–421.

10 HIV Drugs and Sex

10.1 INTRODUCTION

While anybody of any age, gender, or sexual orientation can become infected with HIV, it is most prevalent among adult men who have sex with men (MSM). When the related condition of acquired immune deficiency syndrome (AIDS; Fig. 10.1) was first elucidated in the early 1980s, it was a devastating diagnosis and invariably fatal. Today, the term "AIDS" has been eclipsed by "stage 3 HIV." Advances in medical therapy with novel antiretroviral treatment have changed the trajectory of this condition, to the point that stage 3 HIV is no longer an inevitability, and many people with HIV manage their disease well and go on to live longer, healthier lives. In 1996, the life expectancy of a 20-year-old man with HIV was 39 years, but by 2011, the life expectancy of that same hypothetical person was 70 years. People who are HIV-positive can live healthier and more productive lives than ever before, but that requires a complex antiretroviral regimen. This phenomenon has become so common in healthcare that it has even acquired its own acronym: PLWH (people living with HIV).[1]

The most frequent way that people contract HIV is through sexual encounters. Receptive anal sex is considered the riskiest of all sexual activity, because the lining of the rectum is thin and allows for easier viral entry. For this reason, HIV is most frequent among MSM and others who practice receptive anal sex. Vaginal sex is also a possible route of transmission, and in general such activity is riskier for a woman than a man. Transmission of HIV from oral sex is very low, almost negligible but possible. Sexual activity that does not involve contact with bodily fluids does not promote HIV infection.[2]

10.2 ANTIRETROVIRAL THERAPY AND PREEXPOSURE PROPHYLAXIS

Antiretroviral therapy (ART) typically involves three or more antiretroviral drugs such as nonnucleoside reverse transcriptase inhibitors, nucleoside reverse transcriptase inhibitors, protease inhibitors, entry inhibitors, and integrase inhibitors. With appropriate viral suppression from these drugs, a person may achieve an undetectable viral load, which both improves health and reduces the chances of disease transmission. In fact, the CDC's (Centers for Disease Control and Prevention) position is that maintaining an undetectable viral load effectively eliminates the possibility of sexual transmission. In addition to pharmacological therapy, most people who are HIV-positive under medical care receive education and counseling support.

In a study of 40 men and women with HIV starting ART, a pattern of sexual activity emerged that found diminished sexual desire and less sexual activity at three and six months after starting ART, compared to 18 and 30 months, when sexual desire and activity resumed.[3] This stands in contrast with a popular presumption that ART leads to increased sexual impulsivity and more sexual contact among people who are HIV-positive. ART commencement tends to reduce sexual activity. Many participants abstained from sexual activity for fear of infecting others or out of a desire to adhere to the advice of healthcare professionals or counselors to refrain from sexual activity. The biggest challenge reported by participants in this study was consistent use of condoms.[3] An interesting finding in this study is that many participants preferentially sought prospective sexual partners according

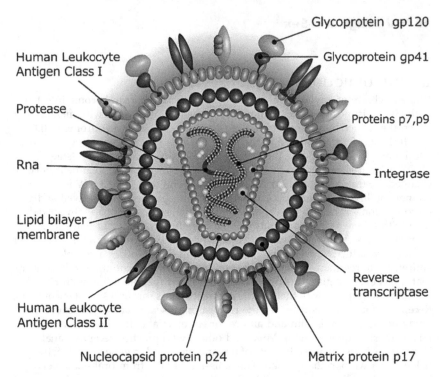

Figure 10.1 An HIV virion.

to their serostatus—that is, people who were HIV-positive sought other people who were HIV-positive for sexual encounters and relationships.[3]

The most frequent route of transmission of HIV is over genital mucosa. Various antiretroviral drugs have been developed, but some have led to HIV-1 reverse transcriptase gene mutations causing drug resistance. Preexposure prophylaxis (PrEP) involves the use of antiretroviral agents in people who are at high risk for HIV but do not yet have the infection. The concept is that use of an antiretroviral agent before infection will reduce the chances of infection, even if the person is exposed. Stampidine, a novel nucleoside reverse transcriptase inhibitor, is a frequently prescribed drug for PrEP.[4]

Because the virus associated with AIDS is a retrovirus (Fig. 10.2), it is by definition an RNA virus with a reverse transcriptase enzyme that allows it to transcribe its RNA into DNA once it invades a host cell. This means the attacked cell then has retroviral DNA, which can then integrate into the chromosomal DNA of the host cell. The antiretroviral drugs developed to fight these viruses inhibit or completely block the enzymes needed to make the RNA-to-DNA switch, but none of them can cure the infection. In addition to antiretrovirals, there are other ART drugs—for instance, entry inhibitors, which block the ability of the HIV virus to enter the host cell.

Postexposure prophylaxis (PEP) involves taking specific medications after a possible HIV exposure. Such an exposure might be due to a broken condom, an errant sexual encounter, or sexual assault. People who seek PEP should get the medication as quickly as possible after the exposure, as it is most effective when

HIV VIRAL LIFE CYCLE

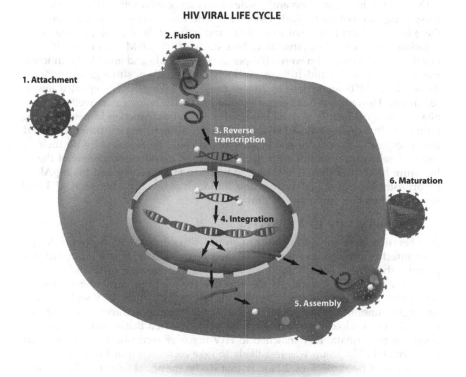

2. Fusion

1. Attachment

3. Reverse transcription

6. Maturation

4. Integration

5. Assembly

Figure 10.2 HIV viral life cycle.

administered immediately. PEP typically involves a 28-day course of antiretroviral therapy.

Treatment as prevention is practiced by people who are HIV-positive who take ART to suppress their viral load to the point that they are at low to no risk of transmitting HIV. The goal of treatment as prevention is to achieve an undetectable viral load.

10.3 SIDE EFFECTS OF HIV DRUGS

There are more thyan 40 antiretroviral agents indicated for HIV, and most people take a cocktail of three or more drugs in combination. Like most medications, these drugs can have side effects that range from mild to severe, and they may interact with other drugs a person takes for other conditions. Some of the most commonly reported side effects associated with HIV therapies are nausea and vomiting, rash, fatigue, loss of appetite, lipodystrophy (gaining weight but only in certain parts of the body), diarrhea, higher cholesterol or triglyceride levels, depression, anxiety, mood changes, and disordered sleep. This list is by no means exhaustive. People taking HIV drugs should not stop them abruptly, even when side effects occur, but should get advice from their healthcare team and rotate to a different agent. Nearly all of these drugs cause rash, fatigue, and nausea or vomiting, but such side effects may resolve after the first few weeks of treatment.

Despite the long list of potential side effects associated with HIV treatment, these drugs are not particularly associated with sexual dysfunction, although there is limited evidence that zalcitabine and enfuvirtide in particular may be associated with erectile dysfunction. In a study of 100 MSM who were HIV-negative and 73 men who were HIV-positive, low libido and erectile dysfunction were reported by 2% and 10%, respectively. In the HIV-positive group, 48% of those taking ART—compared to 26% of those not taking ART—reported lower sex drive.[5] However, it has been speculated that psychological as well as pharmacological factors may be involved in sexual dysfunction in people living with HIV. In a cross-sectional study of 357 men who were HIV-positive, sexual desire, orgasm, and overall sexual satisfaction could be associated with their mental health score but not with ART.[6] However, a systematic review of the literature has found that 51% of people who are HIV-positive and taking ART report some type of sexual dysfunction, although a direct link between ART and sexual dysfunction could not be drawn.[7]

10.4 GROWING UP HIV-POSITIVE

A generation of young adults has grown up with perinatal HIV—that is, they were infected at or around the time of their birth. In 2016, it was estimated that globally there were about 2 million people between the ages of 10 and 19 who were HIV-positive and acquired the infection from a form of mother-to-child transmission.[8] In general, such individuals are aware of their HIV status and undergo counseling and medical care, but as they transition through adolescence, they, like most adolescents, explore their sexuality. In a British study of 392 people with perinatal HIV matched to HIV-negative controls, it was found that the perinatal HIV group was less likely to take recreational or illicit drugs than their negative peers (15% vs. 29%).[9] In that study, the age at first sexual encounter was similar between cohorts, namely 15–19 for the perinatal HIV group compared to a mean of 16 for the HIV-negative group. Risk factors for sexual activity in the HIV-positive group were male sex, poorer quality of life, drug or alcohol use, and older age.[9] About one-third of the HIV-positive group and two-thirds of the HIV-negative group reported not consistently using condoms.[9]

10.5 HIV AMONG SEX WORKERS

The global prevalence of HIV among female sex workers is over 10%.[10] Among sex workers, there are three important subpopulations that are not always evaluated separately in terms of HIV: underage sex workers, male sex workers, and male-to-female transgender sex workers. For a variety of reasons, data on these subpopulations are limited or entirely missing.[11] Sex workers are at very high risk for HIV infection not just because of the nature of their occupation but because many of them face insurmountable barriers to getting adequate HIV preventive care and education.[12] PrEP by definition is intended for at-risk people, but less than 10% of transgender sex workers report having access to it.[12] Furthermore, the stigma against sexual and gender minorities and sex workers is real; even healthcare professionals and social workers are not immune.

In a survey of 181 transgender youth in the United States, 31% were living with HIV infection and 92% reported that they had been tested at least once in their lifetime for HIV.[12] This indicates an awareness of the risk of HIV. In a study of 35 male-to-female transsexual women on ART compared to 2,770 cisgender controls taking ART, the transsexual population was less likely to report adherence rates above 90%, expressed lower levels of confidence about their ability to integrate ART into their day-to-day lives, and reported fewer positive encounters with healthcare providers.[13]

Since sex workers can be a driver of high HIV infection rates, it is surprising that so few studies are conducted in this population. Furthermore, it is regrettable that sex workers concerned about HIV may often experience difficulties in accessing the healthcare system and obtaining ART.[14] Sex work is associated with myriad health problems, including mental health disorders, infertility, and addiction, in addition to HIV.[15] Even when sex workers seek help and support from the healthcare system, they may still face stigma associated with their profession.[16]

10.6 FEMALE-TO-FEMALE TRANSMISSION OF HIV

The risk of female-to-female sexual transmission of HIV is so uncommon that there are only a few known cases. It is much more typical for a lesbian to acquire HIV through blood donation, tattooing, intravenous drug use, sex with a man, or some other route than female-to-female sexual activity. In the few known cases of female-to-female sexual transmission, it has been speculated that transmission involved sex toys.[17–19]

10.7 CONCLUSION

Despite tremendous progress in treating HIV and allowing people who are HIV-positive to live relatively healthy and longer lives, HIV remains a life-altering diagnosis. It can be treated with a cocktail of various antiretroviral and other drugs, and some people on this kind of therapy achieve undetectable viral loads, meaning they are less likely to be able to transmit the disease even during unprotected sex. Antiretroviral and other drugs have mild to severe side effects including nausea and vomiting, fatigue, and rash, but the evidence about their sexual adverse effects is equivocal. Nevertheless, many people who are HIV-positive engage in sexual activity less frequently than they did before their diagnosis. It must be noted that sexual behaviors change with psychological as well as pharmacological factors, and HIV status may cause a person to withdraw from a more active sexual life.

REFERENCES

1. Coelho L, Rebeiro PF, Castilho JL, et al. Early retention in care neither mediates nor modifies the effect of sex and sexual mode of HIV acquisition on HIV survival in the Americas. *AIDS Patient Care STDS*. 2018;32(8):306–313.

2. HIV.gov. Preventing sexual transmission of HIV. https://www.hiv.gov/hiv-basics/hiv-prevention/reducing-sexual-risk/preventing-sexual-transmission-of-hiv. Published 2020. Accessed December 12, 2020.

3. Wamoyi J, Mbonye M, Seeley J, Birungi J, Jaffar S. Changes in sexual desires and behaviours of people living with HIV after initiation of ART: Implications for HIV prevention and health promotion. *BMC Public Health*. 2011;11:633.

4. Uckun FM, Cahn P, Qazi S, D'Cruz O. Stampidine as a promising antiretroviral drug candidate for pre-exposure prophylaxis against sexually transmitted HIV/AIDS. *Expert Opin Investig Drugs*. 2012;21(4):489–500.

5. Lamba H, Goldmeier D, Mackie NE, Scullard G. Antiretroviral therapy is associated with sexual dysfunction and with increased serum oestradiol levels in men. *Int J STD AIDS*. 2004;15(4):234–237.

6. Guaraldi G, Luzi K, Murri R, et al. Sexual dysfunction in HIV-infected men: Role of antiretroviral therapy, hypogonadism and lipodystrophy. *Antivir Ther.* 2007;12(7):1059–1065.

7. Collazos J. Sexual dysfunction in the highly active antiretroviral therapy era. *AIDS Rev.* 2007;9(4):237–245.

8. Slogrove AL, Sohn AH. The global epidemiology of adolescents living with HIV: Time for more granular data to improve adolescent health outcomes. *Curr Opin HIV AIDS.* 2018;13(3):170–178.

9. Judd A, Foster C, Thompson LC, et al. Sexual health of young people with perinatal HIV and HIV negative young people in England. *PLoS ONE.* 2018;13(10):e0205597.

10. Shannon K, Crago AL, Baral SD, et al. The global response and unmet actions for HIV and sex workers. *Lancet.* 2018;392(10148):698–710.

11. Jende JME, Groener JB, Rother C, et al. Association of serum cholesterol levels with peripheral nerve damage in patients with type 2 diabetes. *JAMA Network Open.* 2019;2(5):e194798.

12. Reisner SL, Jadwin-Cakmak L, White Hughto JM, Martinez M, Salomon L, Harper GW. Characterizing the HIV prevention and care continua in a sample of transgender youth in the U.S. *AIDS Behav.* 2017;21(12):3312–3327.

13. Sevelius JM, Carrico A, Johnson MO. Antiretroviral therapy adherence among transgender women living with HIV. *J Assoc Nurses AIDS Care.* 2010;21(3):256–264.

14. Mountain E, Pickles M, Mishra S, Vickerman P, Alary M, Boily MC. The HIV care cascade and antiretroviral therapy in female sex workers: Implications for HIV prevention. *Expert Rev Anti Infect Ther.* 2014;12(10):1203–1219.

15. Ward H, Day S. What happens to women who sell sex? Report of a unique occupational cohort. *Sex Transm Infect.* 2006;82(5):413–417.

16. Benoit C, Jansson SM, Smith M, Flagg J. Prostitution stigma and its effect on the working conditions, personal lives, and health of sex workers. *J Sex Res.* 2018;55(4-5):457–471.

17. Chu SY, Hammett TA, Buehler JW. Update: Epidemiology of reported cases of AIDS in women who report sex only with other women, United States, 1980–1991. *AIDS.* 1992;6(5):518–519.

18. Raiteri R, Fora R, Sinicco A. No HIV-1 transmission through lesbian sex. *Lancet.* 1994;344(8917):270.

19. Chan SK, Thornton LR, Chronister KJ, et al. Likely female-to-female sexual transmission of HIV—Texas, 2012. *MMWR Morb Mortal Wkly Rep.* 2014;63(10):209–212.

11 Performance Drugs (Anabolic-Androgenic Steroids) and Sex

11.1 INTRODUCTION

Anabolic-androgenic steroids are performance-enhancing drugs that have been or are actively being used by about 3 million people in the United States alone.[1] They are Schedule III controlled substances, including oxymetholone, methandrostenolone, stanozolol, nandrolone, and oxandrolone. These agents are synthetic testosterone derivatives and are often taken by athletes to increase muscle mass, improve physical strength, and boost athletic performance (Fig. 11.1). Steroids are available in the United States by prescription; drugs such as glucocorticosteroids can be prescribed for many conditions, including asthma, respiratory dysfunction, and low red blood cell count. Anabolic-androgenic steroids are available by prescription but are not indicated for increasing strength or muscle mass. Obtained illicitly, they are used by amateur and professional athletes and other individuals to bulk up and gain strength. In addition to their being illegal, many sports and athletic programs ban doping, or using such drugs to gain a physical competitive advantage. Athletic competitions and sports may conduct routine and surprise drug tests to determine if any athletes are taking anabolic-androgenic steroids. Newer types of synthetic "designer anabolic steroids" are becoming available on the black market, mainly aimed at providing the muscle-building benefits of anabolic-androgenic steroids in a form that is less detectable or even undetectable in a routine drug test. Illicit anabolic-androgenic steroids may be purchased online, and there are many internet sources for information about these agents, how to take them, and how they should be used.[2]

Patterns of anabolic-androgenic steroid use have been described as cycling (taking multiple doses for a period of time, then stopping cold turkey and restarting again at a later time), stacking (taking two or more types at the same time), pyramiding (slowly building up to a maximum dose over a period of time and then slowly reducing the dose again over time to zero), and plateauing (rotating different steroids or combinations, sometimes overlapping, to prevent tolerance). While the use of anabolic-androgenic steroids may lead to physical dependence, they do not produce a psychoactive high or activate chemical reward circuits in the brain. In other words, anabolic-androgenic steroids are not addictive in the same way that heroin is addictive. In fact, anabolic-androgenic steroid use is associated with few, if any, withdrawal symptoms, and most motivated individuals can stop their use with little problem if they taper off the drugs.[3]

Anabolic-androgenic steroids are synthetic testosterone derivatives whose chemical structures have been altered in such a way that their anabolic or muscle-building effects are enhanced while their androgenic qualities are decreased. The androgenic component of testosterone relates to secondary male sexual characteristics such as a lower voice and facial hair. The anabolic effect promotes increased nitrogen concentration in muscles while preventing catabolic breakdown of muscle tissue. When combined with a systematic and rigorous program of strength training, anabolic-androgenic steroids may increase muscle mass and boost strength as much as 5%–20%.[4] In professional athletics, where tiny differentials can spell the difference between a gold and a bronze medal or between success and failure, this can be a real temptation.

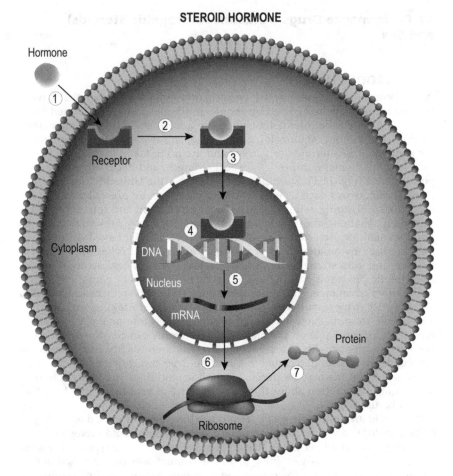

STEROID HORMONE

Figure 11.1 Cell mechanism of steroid hormone action.

11.2 POPULATIONS USING ANABOLIC-ANDROGENIC STEROIDS

While millions around the world use anabolic-androgenic steroids, a few important subpopulations of users emerge.[1] By far the largest group of users are young men who wish to improve their appearance by bulking up or to boost their athletic abilities. These users are young White men (under age 45) with advanced educations and above-average income levels.[5] While most of them play sports or work out, some are elite athletes, such as weightlifters, who find that these drugs give them a significant competitive edge. While better physical performance is the main reason most young White men take anabolic-androgenic steroids, some take them primarily because they feel that being more muscular will make them more attractive. These two motivations can overlap.

Another significant subpopulation of anabolic-androgenic steroid users are (generally) men are over the age of 40, White (93%), and heterosexual (97%), who exercise for sport (79%).[6] When men in this population who do and do not take anabolic-androgenic steroids were compared, the steroid group was more likely to binge-drink (48% vs. 29%) and to have a diagnosed anxiety disorder (12% vs. 3%).[6] Anabolic-androgenic steroid use among older heterosexual men is typically practiced as part of a broader polypharmacy, which may encompass health-related supplements and other drugs (both performance-enhancing and recreational).[6]

Men who have sex with men are increasingly at elevated risk for anabolic-androgenic steroid misuse and must be considered another subpopulation of users. In a survey of 2,733 men who have sex with men in Australia, 5% of respondents said they currently took anabolic-androgenic steroids, but 25% said they were considering starting them.[7] Black sexual-minority boys tend to use these drugs more than Hispanic or White sexual-minority boys (25%, 20%, and 9%, respectively).[8] Among heterosexual boys, the use of anabolic-androgenic steroids was 4%.[8] A study of US adolescent boys found that anabolic-androgenic steroid use was higher among heterosexual boys who were bullied for supposedly being gay or bisexual compared to heterosexual boys who were not bullied in this way.[9]

11.3 SIDE EFFECTS OF ANABOLIC-ANDROGENIC STEROIDS

The side effects of anabolic-androgenic steroids differ by biological sex. Males may develop gynecomastia (breasts), shrunken testicles, an enlarged prostate, and infertility. This seemingly paradoxical effect from taking exogenous synthetic testosterone is likely caused by a combination of testosterone hyper-supplementation and the androgenic effects of the steroids, added to the fact that the use of synthetic testosterone greatly inhibits the body's production of endogenous testosterone. These effects may last for weeks or months after the anabolic-androgenic steroids are stopped. When regular anabolic-androgenic steroid use is halted, the result is an iatrogenic form of low testosterone. In a systematic review ($n = 3,879$ in 31 studies) and a meta-analysis of 11 studies, serum levels of endogenous testosterone were reduced during the course of steroid use and remained reduced for 16 weeks after use was discontinued.[10] The abuse of anabolic-androgenic steroids was also associated with functional and structural changes in sperm and subfertility. Most of the men who were regular users had hypogonadism, which persisted for many weeks or months after use was stopped.[10] Prolonged use of anabolic-androgenic steroids is associated with decreased libido in men.

Females who take anabolic-androgenic steroids may develop menstrual disorders, such as infrequent menstrual periods or amenorrhea. However, extreme exercise alone can also cause amenorrhea, so it is not always clear whether this symptom relates to lifestyle factors or anabolic-androgenic steroid use. Other symptoms of anabolic-androgenic steroid use in females are hirsutism (increased body hair), a lower voice, an enlarged clitoris, and baldness. While most of these side effects resolve when the drug is discontinued, some can be irreversible, such as lower voice and enlarged clitoris. It is not clear why these side effects would persist even long after the drug use has stopped.

Users of any sex may develop acne, tendinitis, tendon rupture, liver dysfunction, hypertension, hypercholesterolemia, and psychiatric disorders such as depression, mania, or delusions. Anabolic-androgenic steroids may make people more prone to infections.[3] Probably the best-known side effect of anabolic-androgenic steroid use is "roid rage," or aggressive, sometimes violent and irrational behaviors which may at times play out in the form of sexual aggression. In its mildest form, roid rage may sharpen competitive drive and encourage athletes to train more aggressively. In more extreme forms, it has been linked to criminal acts, including murder. A definitive link between murder and anabolic-androgenic steroid use often cannot be established, but neither can another coherent motive for the crime.[11] It has been speculated that roid rage is more pronounced in those who had psychological traits of hostility and anger before they took the drug.[3]

11.4 ANABOLIC-ANDROGENIC STEROIDS AND MALE SEXUAL DYSFUNCTION

In a survey of 180 outpatients who took anabolic-androgenic steroids (average age, 34; 99% men), 34% reported that the steroids decreased their libido and 20% reported having erectile dysfunction.[12] In considering this side effect, it is important to understand the role of testosterone in erections. Testosterone does not cause erections; in fact, it is possible to have an erection without testosterone at all. The role of testosterone in sexuality is to help choreograph or synchronize the various physiological responses of male sexual activity, which start with sexual interest or attraction, develop into arousal and erection, and then lead to emission, orgasm, and ejaculation. Testosterone may be considered the synchronizing factor in this cascade, because it regulates enzymes, nitric oxide synthase, and phosphodiesterase type 5 (PDE5), which govern the cycle of erection to detumescence. An erection occurs when nitric oxide (regulated by testosterone), along with other neuroendocrine factors, causes the smooth muscles of the blood vessels to relax and allow the penis to fill with blood, which compresses the veins that drain the corpus cavernosum (the spongy tissue in the penis filled with blood vessels), allowing the erection to be maintained.[13] Testosterone governs both the start and culmination of an erection, so it is not a substance that exclusively causes erections; its actual net effect on the erection-ejaculation cycle is zero.[14] Thus, by interfering with the body's natural endogenous testosterone, anabolic-androgenic steroids can disrupt the ability to properly synchronize erections.

Testosterone can be metabolized in different ways. As an androgen, testosterone can bind to the androgen receptor and reduce to its metabolite, 5-dihydrotestosterone, which is about six times more potent than the parent substance and likewise binds to androgen receptors; it has physiological effects on sexual function, libido, arousal, and orgasm. However, testosterone can also be aromatized to estradiol (estrogen), which can produce effects like fluid retention, adiposity, and gynecomastia. Further complicating the picture, the androgenic aspects of anabolic-androgenic steroids can weaken the ability of testosterone to bind to the androgen receptors. In a study of 321 men taking anabolic-androgenic steroids (80% under the age of 45), a subset ($n = 127$) reported de novo sexual dysfunction (57% lower libido and 27% erectile dysfunction) when they discontinued use. Notably, these symptoms were more severe with longer use and higher doses.[15] This suggests that hyper-supplementation of testosterone using a synthetic product may actually work to suppress the hypothalamic-pituitary-gonadal axis, causing changes in the density of androgen receptors and possibly downregulating endogenous testosterone levels. These effects are masked to a degree as long as a man keeps taking anabolic steroids, but the lower levels of endogenous testosterone are quite apparent when the supplementation stops. Long-term use of anabolic steroids is more associated with sexual dysfunction than short-term or occasional use.

Without the use of anabolic-androgenic steroids or testosterone supplementation, total serum testosterone decreases gradually in men, starting about the third decade of life, so that by about the age of 70, a third of all men have clinically low testosterone (Fig. 11.2).[16,17] This may be considered a natural and expected part of aging. With lower levels of endogenous testosterone, men experience diminished sex drive, erectile dysfunction, and lower energy levels in general.[18] A total testosterone level below 200 ng/dL indicates hypogonadism.[18] When men with low total testosterone levels receive testosterone therapy, sexual function typically improves.[19] Libido, which likewise declines with advancing

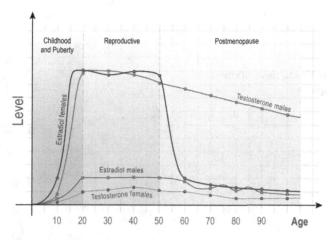

Figure 11.2 Female and male hormone levels as a function of age.

age, can be improved in older men with testosterone treatment.[20] These seemingly paradoxical findings that testosterone can improve sexual function but anabolic-androgenic steroids induce sexual dysfunction warrant explanation. Anabolic-androgenic steroids produce abnormally high levels of testosterone in the body, far higher than the levels in men who have naturally low testosterone levels and receive prescribed testosterone supplementation. It is the abnormally high levels of synthetic testosterone produced by anabolic-androgenic steroids, combined with their androgenic effects and lower endogenous testosterone production, that can cause sexually related symptoms. Androgens play a key role in developing male reproductive organs, and exert their effects at the androgen receptors, which initiate the cascade of physiological events needed for spermatogenesis (Fig. 12.3). To create sperm, high levels of intratesticular testosterone are needed; this testosterone is secreted by Leydig cells.[21] The exact mechanisms of spermatogenesis remain to be elucidated.[22] What is known is that exogenous testosterone creates a negative feedback loop in the hypothalamic-pituitary axis and blocks production of follicular-stimulating hormone and luteinizing hormone needed for sperm production. This can result in oligospermia, azoospermia, and abnormal sperm morphology and motility.[23] This effect is typically reversible but may persist for months, even a year, after anabolic-androgenic steroid use has ceased.[24] Thus, anabolic-androgenic steroid use can interfere with sperm production even to the point of infertility, but fertility may be restored after prolonged abstinence from anabolic-androgenic steroids.[1]

11.5 ANABOLIC STEROIDS AND POLYCYSTIC OVARY SYNDROME

Women who take anabolic-androgenic steroids may have a higher incidence of polycystic ovary syndrome (PCOS). PCOS is really a misnomer, because those with the condition do not necessarily have any ovarian cysts. Instead, it is a catchall term for a variety of different conditions with the common symptoms of irregular menstrual cycles and hirsutism. The underlying cause of PCOS is excess androgen production, which could occur naturally or may occur with androgen supplementation via anabolic-androgenic steroids. PCOS is associated with infertility, obesity, diabetes, heart disease, and premature mortality.

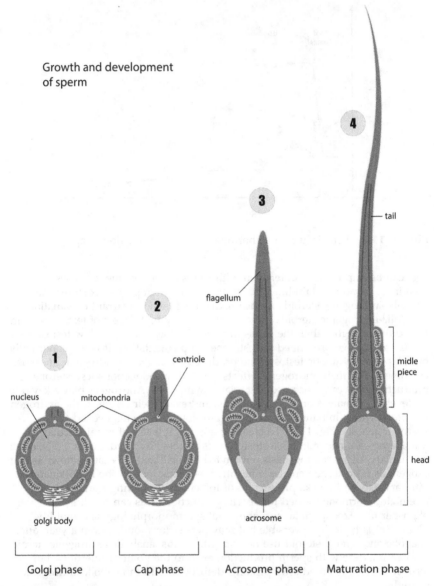

Growth and development of sperm

Figure 11.3 Growth and development of sperm.

Healthy females produce small amounts of endogenous testosterone in the ovaries and adrenal glands. In young premenopausal females, about 75% of testosterone is produced by the adrenal glands and just 25% by the ovaries, but by menopause the balance changes, with the ovaries producing about 50% of testosterone. Ovarian production of estradiol (estrogen) ceases in postmenopausal females, but the ovaries continue to produce androgens for a while. This androgen production likewise diminishes with age. It is the excess androgen provided by anabolic-androgenic steroids that has been associated with PCOS. Women with PCOS may be subfertile or infertile.

11.6 CONCLUSION

Anabolic-androgenic steroids are used not just by elite athletes, but by ordinary persons for a variety of reasons mainly associated with improving performance or developing muscles. These drugs are not benign, and have been associated with a variety of symptoms, including sexual dysfunction. They are not legal for use without a prescription and should only be taken for specific indications under close medical supervision.

REFERENCES

1. El Osta R, Almont T, Diligent C, Hubert N, Eschwège P, Hubert J. Anabolic steroids abuse and male infertility. *Basic Clin Androl.* 2016;26:2.

2. McBride JA, Carson CC, 3rd, Coward RM. The availability and acquisition of illicit anabolic androgenic steroids and testosterone preparations on the internet. *Am J Mens Health.* 2018;12(5):1352–1357.

3. van Amsterdam J, Opperhuizen A, Hartgens F. Adverse health effects of anabolic-androgenic steroids. *Regul Toxicol Pharmacol.* 2010;57(1):117–123.

4. Dandoy C, Gereige RS. Performance-enhancing drugs. *Pediatr Rev.* 2012;33(6):265–271; quiz 271-262.

5. Cohen J, Collins R, Darkes J, Gwartney D. A league of their own: Demographics, motivations and patterns of use of 1,955 male adult non-medical anabolic steroid users in the United States. *J Int Soc Sports Nutr.* 2007;4:12.

6. Ip EJ, Trinh K, Tenerowicz MJ, Pal J, Lindfelt TA, Perry PJ. Characteristics and behaviors of older male anabolic steroid users. *J Pharm Pract.* 2015;28(5):450–456.

7. Griffiths S, Murray SB, Dunn M, Blashill AJ. Anabolic steroid use among gay and bisexual men living in Australia and New Zealand: Associations with demographics, body dissatisfaction, eating disorder psychopathology, and quality of life. *Drug Alcohol Depend.* 2017;181:170–176.

8. Blashill AJ, Calzo JP, Griffiths S, Murray SB. Anabolic steroid misuse among US adolescent boys: Disparities by sexual orientation and race/ethnicity. *Am J Public Health.* 2017;107(2):319–321.

9. Parent MC, Bradstreet TC. Sexual orientation, bullying for being labeled gay or bisexual, and steroid use among US adolescent boys. *J Health Psychol.* 2018;23(4):608–617.

10. Christou MA, Christou PA, Markozannes G, Tsatsoulis A, Mastorakos G, Tigas S. Effects of anabolic androgenic steroids on the reproductive system of athletes and recreational users: A systematic review and meta-analysis. *Sports Med.* 2017;47(9):1869–1883.

11. Beresford T. These people were hopped up on steroids—and brutally murdered people. *Ranker*. https://www.ranker.com/list/steroid-murders/trilby-beresford. Published 2017. Accessed December 14, 2020.

12. Smit DL, de Ronde W. Outpatient clinic for users of anabolic androgenic steroids: An overview. *Neth J Med*. 2018;76(4):167.

13. Andersson KE. Mechanisms of penile erection and basis for pharmacological treatment of erectile dysfunction. *Pharmacol Rev*. 2011;63(4):811–859.

14. Vignozzi L, Corona G, Petrone L, et al. Testosterone and sexual activity. *J Endocrinol Invest*. 2005;28(3 Suppl):39–44.

15. Armstrong JM, Avant RA, Charchenko CM, et al. Impact of anabolic androgenic steroids on sexual function. *Transl Androl Urol*. 2018;7(3):483–489.

16. Harman SM, Metter EJ, Tobin JD, Pearson J, Blackman MR. Longitudinal effects of aging on serum total and free testosterone levels in healthy men. Baltimore longitudinal study of aging. *J Clin Endocrinol Metab*. 2001;86(2):724–731.

17. Feldman HA, Longcope C, Derby CA, et al. Age trends in the level of serum testosterone and other hormones in middle-aged men: Longitudinal results from the Massachusetts male aging study. *J Clin Endocrinol Metab*. 2002;87(2):589–598.

18. Schubert M, Jockenhövel F. Late-onset hypogonadism in the aging male (LOH): Definition, diagnostic and clinical aspects. *J Endocrinol Invest*. 2005;28(3 Suppl):23–27.

19. Chiang HS, Cho SL, Lin YC, Hwang TI. Testosterone gel monotherapy improves sexual function of hypogonadal men mainly through restoring erection: Evaluation by IIEF score. *Urology*. 2009;73(4):762–766.

20. Corona G, Isidori AM, Buvat J, et al. Testosterone supplementation and sexual function: A meta-analysis study. *J Sex Med*. 2014;11(6):1577–1592.

21. Sultan C, Gobinet J, Terouanne B, et al. The androgen receptor: Molecular pathology. *J Soc Biol*. 2002;196(3):223–240.

22. Verhoeven G, Willems A, Denolet E, Swinnen JV, De Gendt K. Androgens and spermatogenesis: Lessons from transgenic mouse models. *Philos Trans R Soc Lond B Biol Sci*. 2010;365(1546):1537–1556.

23. Dohle GR, Smit M, Weber RF. Androgens and male fertility. *World J Urol*. 2003;21(5):341–345.

24. Knuth UA, Maniera H, Nieschlag E. Anabolic steroids and semen parameters in bodybuilders. *Fertil Steril*. 1989;52(6):1041–1047.

12 Medical Devices and Alcohol

12.1 INTRODUCTION

While this book is devoted to pharmacological products, sexual side effects have been reported with medical devices, such as pacemakers and implanted defibrillators, and with alcohol—which is not a pharmaceutical product but has the properties of a drug. In the interest of being complete and addressing related questions of medical interest, we include some observations.

12.2 IMPLANTABLE CARDIOVERTER-DEFIBRILLATORS AND PACEMAKERS

Implantable cardioverter-defibrillators (ICDs) are pacemaker-like devices (Fig. 12.1) that rescue people from potentially life-threatening arrhythmias by delivering a high-energy shock to "reset" the heart by sudden and total depolarization. While "therapy delivery," as it is euphemistically described, can be a life-saving intervention, this high-voltage shock often occurs suddenly and without warning, and has been described by some patients as like being kicked in the chest by a mule. ICD shocks can be upsetting, surprising, painful, and psychologically distressing. People who get an ICD shock often confront the notion that they just had a potentially fatal arrhythmia. While being saved from death is a good thing, some people with ICDs take these shocks as grim reminders of their own mortality. To that end, ICD support groups are offered at many hospitals to help people with these implantable devices come to terms with what it means to live with one. Sexual activity is a topic these groups sometimes address.

In a survey of 443 US people with ICDs, 65% reported that they were able to engage in sexual activity, but 51% said they chose not to have sex for a variety of reasons. Among those reasons were fear that the sex would trigger a shock by the ICD, worry that the sex might cause the heart to beat too rapidly, lack of desire, or doctor's orders to avoid sex.[1] It is important to note in this context that implantation of an ICD may have been preceded by a serious cardiac event; some people may have survived a near-death experience.[2] Most ICD recipients never experience a shock, but among those who do, some develop an aversion or even responses resembling posttraumatic stress.[2] In a study of 308 ICD recipients, it was found that the psychological distress that occurs around the time of implant subsides with time,[3] and some recipients even find that the implanted device gives them a sense of renewed safety and assurance. They tend to view the device positively and see life after implantation as a second chance.

Technically, it is possible for an ICD recipient to receive a shock at any time, and about 7%–13% of recipients who have regular sex will experience shock therapy during a sexual encounter. While the ICD recipient feels the shock, it is not transmitted through the body to the partner. The fear of dying during sexual activity occurs in recipients and their partners. About 13% of recipients are afraid that they might die during sex; about 26% of their partners are afraid the recipient might die during sex.[4]

Most ICD recipients are older individuals with heart disease that predates the implantation. In a survey of 415 men with ICDs (average age, 64), 70% had erectile dysfunction, 58% had orgasmic dysfunction, 83% had diminished libido, and 77% had decreased overall sexual satisfaction.[5] However, it must be noted

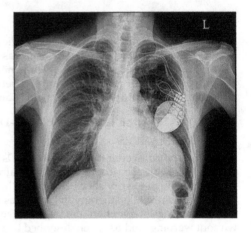

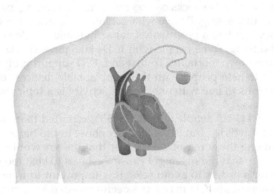

Figure 12.1 Implantable cardioverter-defibrillator and placement.

that erectile dysfunction is highly prevalent among men with ischemic heart disease and heart failure, two of the main indications for an ICD. Sex in ICD recipients has not been well studied, and we were unable to find any reports in the literature of an ICD shock causing death or injury in a person engaging in sexual activity.

For people with pacemakers, there is less concern. Pacemakers can only deliver low-voltage stimulation to the heart, typically around 1–2 V. People with pacemakers do not typically feel the pacing support that the device provides. While sex may cause elevated heart rates for people with pacemakers, it is typically about the same increase experienced by going up two flights of stairs, and the pacemaker is unable to deliver any high-energy shocks. While people with pacemakers may experience some of the same initial psychological distress at needing an implantable device for a heart condition, there is no shock or unpleasant device-related effect that could happen during sex.

12.3 INTRATHECAL DRUG PUMPS

Intrathecal drug pumps are implantable devices that deliver a small amount of pain medication (sometimes baclofen) directly to the intrathecal space around the spine. The benefit of this implantable system is that the benefits of drug therapy can be obtained with only small amounts of the drug, because of how it is delivered. However, recipients of these implanted pumps should postpone sexual activity for several weeks after implantation and then seek advice from a physician about if and when sexual activity may resume. Many people who are implanted with these pumps have severe symptoms and already abstain from sexual activity. In a survey of 36 people with an intrathecal drug pump for pain control, the device had no effect on sexual activity; those who had sex before implantation continued to have it afterward.[6]

12.4 TRANSCUTANEOUS ELECTRICAL NERVE STIMULATORS

Transcutaneous electrical nerve stimulation (TENS; Fig. 12.2) uses a device that delivers low-voltage current through electrodes adhered to the skin in order to help reduce painful symptoms. TENS devices are typically used by people with back pain or other conditions. They are also sometimes used during the early stages of labor to provide pain control and stimulate the production of endorphins, natural pain relievers in the body. TENS devices are largely under the control of the patient, who can move the stick-on electrode pads as desired to get maximum benefit. People who use TENS units for pain control do not use them all of the time, so sexual activity is not particularly affected.

Electrosex, or electrical play, is a specific type of sexual activity that uses TENS devices for sexual pleasure rather than pain relief. In such cases, TENS units (which can be purchased without prescription and cost under $100) can be used for self-stimulation by an individual or that individual can give

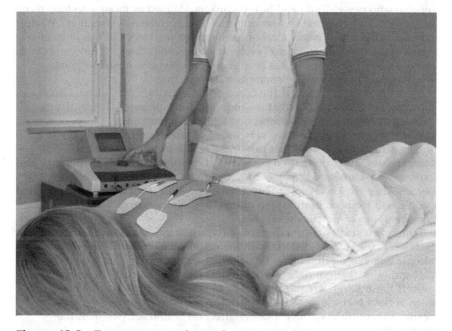

Figure 12.2 Transcutaneous electrical nerve stimulation.

electrostimulation to the partner. Proper placement of the pads can allow the TENS device to deliver enough electrical energy to cause certain muscle contractions that can be perceived as erotic or orgasmic or painful, depending on the location. TENS electrosex should not be used in people with an implanted pacemaker or defibrillator (the devices can interfere with each other), in people with heart disease or epilepsy, or in those who are pregnant. Furthermore, TENS devices should not be used around the face, neck, or head.

12.5 ALCOHOL

Alcohol has paradoxical effects on sex. On the one hand, drinking alcohol can lower inhibitions, improve confidence, and even enhance sexual desire. When consumed within reasonable amounts, alcohol may increase libido and inspire sexual activity. Alcohol may make men more sociable and more responsive to a wider range of partners (even those who are "less attractive") and give them more confidence to flirt and make sexual advances. However, alcohol is associated with sexual dysfunction as well, particularly in people who drink to excess or who are alcoholic. Alcohol intoxication can result in erectile dysfunction and anorgasmia. Moreover, alcohol intoxication can lead to risky sexual behaviors, an inability to consent to sex, and sexual regret.

Less is known about the combination of alcohol and other drugs in terms of how they affect sexual behaviors. Many people who use recreational drugs add alcohol if it is available. It is not known how various mixtures (alcohol plus marijuana, alcohol plus cocaine, alcohol plus methamphetamines, and so on) affect libido and sexual performance.[7]

In a meta-analysis, light to moderate consumption of alcohol, defined as fewer than 21 drinks per week, was associated with a decreased risk of erectile dysfunction (odds ratio, 0.71; 95% confidence interval, 0.59–0.86) but teetotalers (people who never drink) and heavy drinkers (defined as those who took more than 21 drinks per week) had no association with erectile dysfunction.[8] While popular belief is that heavy alcohol consumption increases the risk of erectile dysfunction, the epidemiological evidence is less compelling.[9,10] Among women, heavy alcohol consumption has been shown to attenuate sexual response, whereas light to moderate consumption has not.[11]

"Alcohol myopia" refers to the concept that intoxication by alcohol decreases the amount of information that the individual can process. That is, drunk people pay less attention to everything in their surroundings.[12] Alcohol myopia explains why alcohol can have very different effects on the same person in similar situations, because that person bases their responses on the social context, cues, and information perceived at the time. Alcohol myopia may also explain why intoxicated people may engage in risky or unprotected sex, because they are too myopic to consider the possible dangers—most intoxicated people will focus on gratification of their sexual urges over the more serious considerations of practicing safe sex.[13]

Drinking a small amount of alcohol (typically about two drinks) may boost testosterone levels, associated with increased sexual desire. Furthermore, alcohol enjoys a reputation for being an important part of romantic endeavors, promoting relaxation, reducing inhibitions, and making sex more enjoyable, even if these things are not buttressed by strong empirical evidence. It has been argued that the expectation that alcohol is an important prelude to sexual adventure may have a larger effect on sexual activity than the alcohol itself.[14]

REFERENCES

1. Cutitta KE, Woodrow LK, Ford J, et al. Shocktivity: Ability and avoidance of daily activity behaviors in ICD patients. *J Cardiopulm Rehabil Prev.* 2014;34(4):241–247.

2. Bostwick JM, Sola CL. An updated review of implantable cardioverter/defibrillators, induced anxiety, and quality of life. *Psychiatr Clin North Am.* 2007;30(4):677–688.

3. Kapa S, Rotondi-Trevisan D, Mariano Z, et al. Psychopathology in patients with ICDs over time: Results of a prospective study. *Pacing Clin Electrophysiol.* 2010;33(2):198–208.

4. Thompson D. Sex is safe for heart patients with a defibrillator. *WebMD.* https://www.webmd.com/heart-disease/atrial-fibrillation/news/20151109/sex-is-safe-for-heart-patients-with-a-defibrillator#:~:text=An%20ICD%20shock%20during%20sex,patients%20with%20such%20an%20implant. Published 2015. Accessed December 13, 2020.

5. Palm P, Zwisler AD, Svendsen JH, Giraldi A, Rasmussen ML, Berg SK. Compromised sexual health among male patients with implantable cardioverter defibrillator: A cross-sectional questionnaire study. *Sex Med.* 2019;7(2):169–176.

6. Raphael JH, Southall JL, Gnanadurai TV, Treharne GJ, Kitas GD. Long-term experience with implanted intrathecal drug administration systems for failed back syndrome and chronic mechanical low back pain. *BMC Musculosk Disorders.* 2002;3(1):17.

7. Peugh J, Belenko S. Alcohol, drugs and sexual function: A review. *J Psychoactive Drugs.* 2001;33(3):223–232.

8. Wang XM, Bai YJ, Yang YB, Li JH, Tang Y, Han P. Alcohol intake and risk of erectile dysfunction: A dose-response meta-analysis of observational studies. *Int J Impot Res.* 2018;30(6):342–351.

9. Cheng JY, Ng EM, Chen RY, Ko JS. Alcohol consumption and erectile dysfunction: Meta-analysis of population-based studies. *Int J Impot Res.* 2007;19(4):343–352.

10. George WH, Cue Davis K, Schraufnagel TJ, et al. Later that night: Descending alcohol intoxication and men's sexual arousal. *Am J Mens Health.* 2008;2(1):76–86.

11. George WH, Davis KC, Heiman JR, et al. Women's sexual arousal: Effects of high alcohol dosages and self-control instructions. *Horm Behav.* 2011;59(5):730–738.

12. Sevincer AT, Oettingen G. Alcohol myopia and goal commitment. *Front Psychol.* 2014;5:169.

13. Griffin JA, Umstattd MR, Usdan SL. Alcohol use and high-risk sexual behavior among collegiate women: A review of research on alcohol myopia theory. *J Am Coll Health*. 2010;58(6):523–532.

14. Patrick ME, Maggs JL. Does drinking lead to sex? Daily alcohol-sex behaviors and expectancies among college students. *Psychol Addict Behav*. 2009;23(3):472–481.

Index

Printed in the United States
by Baker & Taylor Publisher Services

Printed in the United States
by Baker & Taylor Publisher Services